History of Science and Technology

History of Science and Technology

Exploring New Themes

Edited by

Rattan Lal Hangloo

RAWAT PUBLICATIONS

Jaipur • New Delhi • Bangalore • Hyderabad • Guwahati

ISBN 978-81-316-0426-7

Published by
Prem Rawat for **Rawat Publications**
Satyam Apts, Sector 3, Jawahar Nagar, Jaipur 302 004 (India)
Phone: 0141 265 1748 / 7006 Fax: 0141 265 1748
E-mail: info@rawatbooks.com
Website: www.rawatbooks.com

New Delhi Office
4858/24, Ansari Road, Daryaganj, New Delhi 110 002
Phone: 011 2326 3290

Also at *Bangalore*, *Hyderabad* and *Guwahati*

Typeset by Rawat Computers, Jaipur
Printed at Chaman Enterprises, New Delhi

Dedicated to

an eminent scientist and educationist

Professor Goverdhan Mehta Ji

former Vice-Chancellor, University of Hyderabad
and former Director, IISc, Bangalore

Contents

Preface

The history of science and technology in India is an important subject and requires far more space and serious attention than would be available here. Recent developments in science and technology and their application in social sciences have led us to a situation where there is a rethinking on all the basic concepts and fundamental issues. There is hardly any area that is not impacted by the new developments. At the same time, its rapidity and scale are also too difficult to comprehend and like science fiction, with newer developments coming by the day, even recent advances seem old. Many historians have already started to comprehend these developments and are using them in their research works to bring out new findings which were not only laborious till yesterday but also not precise. The fact, however, remains that many of these advances are still restricted to specialists in their field and they have domain knowledge in their particular area. Under these conditions, an umbrella that discussed these advances under a thematic unity was a well considered move and hence a conference (September 2005) was devoted to this aspect. I owe a very special thanks to all the contributors who devoted adequate time to their respective themes and participated in the national seminar with full texts of their papers, vibrant discussions, exacting criticism and raised very important questions regarding how the researches for reconstructing the history of science in India should proceed.

The persons who were instrumental in motivating me to take up this task of bringing together the scholars working in this area are our country's four eminent scientists – Professor Goverdhan Mehta, Professor Palle Rama Rao, Professor Kota Harinarayana and Professor Syed Ehtisham Hasnain – all were Vice Chancellors of the

University of Hyderabad, one after the other. I am very grateful to them for the multitude of ways in which they provided support and encouragement.

I am very grateful to University of Hyderabad for providing financial support under University with Potential for Excellence (UPE) Programme which made this national seminar possible. I owe a deep debt of gratitude to the Professor V. Kannan, the Pro-Vice Chancellor for his much needed guidance and support. I also thank my colleagues for their unswerving support and cooperation at every step. I also acknowledge valuable inputs by my colleagues Dr. M.N. Rajesh, Professor P. Appa Rao and Professor A. Murali, Department of History, University of Hyderabad.

A special note of appreciation must go to some of my research scholars, particularly Professor Victor Babu, Head, Department of History, Lucknow Central University; Dr. Salma Ahmad Farooqui, Associate Professor and Head, Department of History, Moulana Azad National Urdu University, Hyderabad; Salil Kader, Cultural Specialist, American Consulate at Hyderabad; Senthil Kumar, Officer in American Consulate, Hyderabad; Yaseer Arafat, Assistant Professor, Department of History, Delhi University; V. Dileep Kumar, Assistant Archivist in the Department of History, Hyderabad University and our students, specially M. Ramakrishna and Phenish Babu, Department of History, University of Hyderabad for meticulously attending to every task associated with this national seminar from beginning to end. I also thank Dr. Ajaya Kumar Sahoo of Centre for the Study of Indian Diaspora, University of Hyderabad for his valuable suggestions. My friends Professor Harbans Mukhia, Professor V. Ramakrishna, Dr. Laxminarayana of the Department of Sociology, Dr. Gautam Sengupta, Director General Archaeology Survey of India, Professor Paramjit Singh Ramana, and Shri G.S. Sra deserve special thanks for their encouragement. I am also grateful to Shri Pranit Rawat of Rawat Publications for his support and cooperation in the publication of this book.

Finally, I am grateful to my wife Sharika Kaul who made my devotion to this effort more meaningful and interesting than my other non-academic responsibilities in the University.

Rattan Lal Hangloo

1 Introduction

Rattan Lal Hangloo

There has been a relook at Indian history since the 1990s, which is marked by a profusion of new works. Many of them have adopted refreshing ways of looking at the past and questioned number of basic assumptions. Keeping these advances in mind, a conference on exploring the history of science and technology in Indian history was envisaged to bring together established scholars and young researchers under the thematic umbrella of history of science and technology.

In its search for a holistic understanding of the life cycle of the past societies, archaeology, like many other disciplines, is turning to the physical and biological sciences, facilitating in the process, a convergence of disciplines. From the better known methods of dating, archaeology has now moved on to issues like the levels of technology, choice of crops, dietary habits etc. Archaeological science has emerged as a discipline on its own strength and has enlarged the horizon of the history of science and technology.

Although the history of science and technology is a relatively new entrant to the domain of history, it has attracted a substantial number of researchers into its fold. More important, it is now being taught as an important component of the postgraduate course in history in most Indian universities. One can compare the enthusiasm about the history of science and technology now with the rather lukewarm of response to the discipline in 70s of the last century. Not only has the discipline acquired a large number of votaries, there has been a qualitative leap in the issues addressed by the researchers such as Deepak Kumar and Druv Raina. Also, there have been innovations in research methodologies, an increasing convergence of various branches of science, humanities and social science and above all an explosion of research publications in this area. The practitioners of the discipline now stand on firm ground.

Beginning with the late Professor D.P. Chattopadhyay's pioneering critical work on the philosophical basis of India's scientific temper to his edited book *History of Science, Philosophy and Culture in Indian Civilization* which runs into many volumes – the discipline of the history of science and technology has indeed taken several twists and turns by expanding its frontier, interacting with a number of other disciplines and opening up new areas of enquiry. This journey had not always been smooth. Like many other areas of enquiries, the search for an objective understanding of the intellectual roots of pre-modern science in India has often degenerated into chauvinistic, politicised statement of a glorification of the past. The search for an authentic Indian science has chosen to overlook, often purposefully, the pluralistic tradition of the subcontinent, sharing and borrowing ideas and practices. Again, there had been subtle and not-so-subtle attempts at viewing 'science' in pre-modern or more specifically ancient India in terms of monolithic Sanskrit texts and near total rejection of the richness and strength of non-textual, multi-layered popular tradition. At times, such attempts turned out to be ridiculous and were promptly dismissed from academic space, but alarmingly some such acts of distortions have come to stay and even gained legitimacy through repeated pronouncement. Another important aspect of the research in these disciplines is the increasing awareness about the relevance of the ethnic traditions of technology as explanatory tools.

The interface with many other disciplines like field archaeology, physical anthropology, metallurgy, history of science and religion, defence technology, statistics, trade, craft technology and other related areas are an integral part of the articles. The main concern is to see how methods from these disciplines can be applied to the study of Indian history. Therefore this volume represents an attempt to provide an introduction to the variety of aspects that are being explored by a number of scholars from various regions cutting across time and space dimension in Indian history.

Arun Sonakia's article "Narmada *Homo Erectus*: The Most Primitive Human Fossil of India" based on fieldwork and discovery of the skeleton in India, has pushed back the dawn of human settlement by a few thousand years ago and also provided us with a sheet anchor to locate the same find in a temporal and spatial context. The significance of this article lies in the fact that this is the only human fossil that bridges the gap between Africa, Europe and

West Asia on side one and China and Indonesia on the other on a global map of human fossils. The megaliths of peninsular and southern India represent one of the enigmatic aspects of Indian archaeology, with a few burial-cum-habitation sites having been uncovered near cairns or burials.

The article of Sharada Srinivasan "Megalithic High Tin Bronzes" explores important aspects of technology related to archaeological issues of this intriguing material culture. Archaeo-metallurgical investigations by Sharada Srinivasan, on a range of thin vessel fragments from a range of megalithic sites such as Adichanallur and Nilgiris in Tamil Nadu, and Mahurjhari in the Vidharbha region, point significantly to the use of binary copper tin alloys with a higher content of tin in the sites of Nilgiris and Adichanallur in particular, showing the well developed use of wrought and quenched high tin beta bronze (with around 23% tin).

Views regarding subsistence strategies of the Iron Age in peninsular India have remained consistently divided. These have ranged from a nomadic, pastoral way of life to people who practised agriculture and were responsible for introducing irrigation to South India. Oddly enough, little emphasis has been given to the technologies adopted to produce the varieties of objects used in the Iron Age.

Jaya Menon's article "Crafts on the Move? The Case of Megaliths of Peninsular India" attempts to investigate the crafts of the Iron Age in peninsular India and their viability to be produced by nomadic or sedentary people.

The post-Mauryan period in Indian history can be viewed as a period which saw the emergence and gradual crystallisation of contacts between centres of power in the north-west and other regions of the subcontinent. There was an impressive movement of men and merchandise resulting in exchange of ideas and knowledge. Interaction of one kind or the other between the Indo-Iranian borderlands (the river Indus forms the eastern boundary and the western border is defined by the highlands of the Iranian plateau; it is virtually the only overland access to the Indian subcontinent from the west) and the Doab became more intense as the major urban centres of the period like Taxila in the north-west and Mathura in the Doab were located on the great trade route that linked the Gangetic region with the north-west.

Suchandra Ghosh in her article "In Search of Technology: Journey from Ai Khanum to Mathura via Taxila (200 BC–AD 300)" illustrates another important archaeological site called Ai Khanum in Afghanistan which had linkages with Taxila and other places further east. She has shown how common political control during 1st century BC to 3rd century AD in this area created such a situation that the productive forces achieved the highest level of development. Her article is based on the archaeological findings from Taxila and Mathura which demonstrates a very rich material culture. She states that minor objects like beads, utensils, jewellery, tools, glass objects etc., found from these sites betrayed examples of technological understanding in their own way. Interestingly, some texts of the period in question indicate a society where people made the best use of their technological know-how. This article also surveys the technological situation, particularly relating to things of everyday use, of some areas beginning from Ai Khanum in Afghanistan to Mathura in the Doab, with Taxila as the central point between the east and west in the post-Mauryan period.

Mathura is distinct from the other localities of the Ganga valley as it became an integral part in the political changes that were affecting north-western and western India from the second century BC. Right from the rule of the Indo-Greeks to the Kushanas, Mathura was always connected to the then centre of political gravity, i.e., the north-west and was also linked with the Delhi–Aravalli axis and the Cambay node.

The archaeological material, substantiated and supplemented by literary references, indicate remarkable technological know-how relating to chemical practices, metallurgy, metal working, production and colouring of glass, manufacturing of ceramic wares, making of weapons, agricultural implements, domestic utensils, jewellery and much more. This was perhaps possible because of free movement of people, trade and ideas. Another important factor behind technological advancement in the area reviewed in this article is definitely its multipronged linkages of which the central Asian connection played a significant role. With a mélange of archaeology and text, significant attempt has been made in this article to situate the technological tradition, relating particularly to minor objects, of the period and the geographical locale under review through the understanding of a practitioner of history.

Metal craft had formed a part of the rural and tribal scene in Bengal since early times. Historical data bears this out. History has recorded the classical tradition of image crafting more meticulously so that we know about the Pala–Sena bronzes, while there is no record of the beginning of the tribal castings. Most of the *Pala* metal images fit into a discernible group of craftsmanship and metal components. The Pala–Sena bronze images have been objects of serious and interesting studies.

The eastern Indian tradition of image crafting had extended up to Jhewari in Chattagram district of the present Bangladesh, in the east. The main centres where metal image crafting flourished were Nalanda and the Kurkihar, Buddhist sites in Bihar, while, Banpur in Puri district, Orissa formed the southern extension where the tradition could be located. In the medieval times, Nadia and Murshidabad had developed into cultural centres, where minor metal image making thrived. On the other hand minor metal utensils and accessories became more and more popular in the medieval times. Along with Nadia and Murshidabad, Burdwan and Bankura also became centres of production.

Forts have played a major role in the political process of Deccan during the medieval period and the this region has one of the largest concentration of forts in the whole of the country. Departing from the conventional view of studying forts from the perspective of art history or political history, M.N. Rajesh and M. Ramakrishna's article titled "Architecture and Technological Advancement: A Case Study of Select Forts of Deccan" tries to locate the role of forts in the changing context of technology. Technological changes were immediately perceptible in warfare and led to large-scale changes in the organisation of the army and the redrawing of the political boundaries as seen in the rise and fall of many polities. Along with the building of forts, one also sees the absorption of technology in the constant alteration of the structures where new forts were built or altered to incorporate these technologies. Many of them are also found in the additional structures as seen in the forts of Bidar, Gulburga and Bellary etc.

In her article, "A Social History of a Minor Technology: The Tradition of Bronzes, Brasseri Bell Metaland Dokra Manufacture in Bengal", Nupur Dasgupta has examined the subject of minor metal works practised in rural, suburban and tribal environs in the state of West Bengal within a historical context. She has categorised the products into two broad components: (a) effegies/figurines/cast images, and (b) utensils. Of these the first category is marked by two distinct products, rural/urban and tribal. The temporal context begins with the medieval period and continues to the present. She has evolved a theory of continuity or evolution of the monolithic technology and has significantly contributed to a composite perspective of a continued tradition. She has illustrated the story of *Dokra* as another genre and clearly shown how it developed within tribal society and came into notice only with the nineteenth century colonial forays into anthropological studies in the eastern and central India. The tradition of image crafting was kept alive by private worshippers in the medieval and pre-modern times. The 19th century survey reports contain valuable data favourable for ethnological reconstruction. A brief survey of the entire gamut of the minor metal craft from medieval times onwards, reveals a complex picture with multiple stylistic, technical and social paradigms. A thorough study would need a lot of time and a multidisciplinary approach.

Given the situation, Nupur Dasgupta has presented a resume of the kind of data that is available on this subject and a report on the possibilities of research into the history of the minor metal product manufacturing technology in Bengal from the medieval to the present times. The contemporary data in this article is especially handy in tracing aspects of the age-old traditions of manufacture. One very pertinent point to be noted here is that the history of the craft cannot be pursued if we try to separate the craft or products from the social genre, which is producing this craft. She also approaches the history of this technology from a social history perspective, where the craftsmen are as important as their craft and seen as the prime movers in the story of this craft tradition through the ages. The tribal section live in a situation very different than their rural counterparts, and the rural from their suburban brethren. These multiple contexts are reflected in their technology, aesthetics and commercial ventures. The article contains a review of all these aspects in brief based on the data collected so far. The article is based on a project work assisted by Pranab Chattopadhyay and Banani Bhattacharyya.

The article by Vijaya Ramaswamy looks at a much neglected area, the north-west frontier of India in the early historic period when there was a brisk contact with the Greek world. The article entitled "An Overview of Craft Technologies and Craft Communities in Peninsular India: Sixteenth to Eighteenth Centuries" highlights the socio-cultural matrix of the artisans and also how the societal basis for science was created and nurtured by the artisan communities. The study of crafts in pre-colonial peninsular India can be problematised in terms of four specific areas of analysis. These rubrics are:

(a) Crafts and the Artisan Community;
(b) Crafts and Caste;
(c) Technology, Crafts and Traditional Beliefs; and
(d) Gender Issues in Crafts.

She has pointed out the very significant aspect of how the study of this crucial phase of South Indian history has given rise to a number of theories and an equal number of questions. Her article defines how the political and socio-economic milieu of the pre-colonial era provides the context for this study of technology, society and traditional crafts.

In Rattan Lal Hangloo's article "Glimpses of Craft Technology in Kashmir" the narrative travels through different specimens that signify various typologies and also establishes strong assertions about the technological flow and their received understanding in Kashmir. The fact that the stone architecture models as exemplified in the plinths and mouldings had a long historical past is clearly shown in the earlier Buddhist and Hindu models. This was carried forward and improvised during the 13th century and later but is an extension of a practice long held that the Kashmiris were familiar with. Similarly the use of the doorways and wood is something essential to understand the aspect of mosque and Islamic architecture in Kashmir. Large-scale buildings under the patronage of Zainul Abidin signify the clear link between the patronage and the production of new forms. These varieties were from diverse influences from as far as Central Asia and are exemplified in architecture, papier-mache, inlay, engravings, shawls, carpet weaves and a long list of crafts that are uniquely Kashmiri.

K.N. Sethi's article "Textile Technology in Medieval Orissa: A Case Study of Production and Export during Seventeenth and Eighteenth Centuries" shows how production and distribution of economic goods Orissa was based on the co-existence and, at times, inter-penetration of subsistence and a commercialised sector. This article studies how the bulk of the population lived in the villages and how their needs for goods and services satisfied through production for use and a network of reciprocal obligations and exchange accounted for a relatively small proportion of economic activity. Yet exchange of goods, found at virtually every level and sphere of economic life, was impressive in its magnitude and complexity.

Syed Ali Nadeem Rezavi's article entitled "Hydraulics and Water Management at Fatehpur Sikri" recounts the transformation of Fatehpur as a capital city from the village of Sikri was a "well-watered" ground which could sustain a large army. He reveals that the lake of Fatehpur Sikri remained the major source of water supply to the city. "To supply the city with water a tank has been carefully and labouriously constructed two miles long and half a mile wide". This rivulet provided water to the area situated to the east of Fatehpur Sikri. The demand for potable and irrigation water was also taken care of by digging a large number of wells (*chah)* and constructing a number of *baolis* (step-wells) scattered all over the town. He also focuses on the main issue of how water harvesting appears to have been the major objective of the tanks situated on the top of the ridge. The major source of water for these tanks was rain water which was harnessed into them through slopes and channels. The water thus supplied from the northern and southern waterworks to the imperial complex was distributed to various sections through the conducts running between the wells. Probably the entire needs of the eastern area were met through the water stored in this tank.

In his article titled "Air Dynamics and Building Construction in Medieval India", S.S. Subodh has tried to illustrate the nature and role of air dynamics in building construction of the 'Sheesh Mahal' at Farrukhnagar. He has studied Sheesh Mahal not only because of its historical importance but more so because of the architectural specificity among the medieval monuments. It draws the attention of a historian to this architectural specificity that has a very important relationship with the climatology in medieval times.

In a similar vein numismatics and the relationship between the flow of wealth and the metallurgical techniques under the great Mughals is analysed by Najaf Haider in his very important article, "Minting Technology in Mughal India".

Continuing the pragmatic side was R. Balasubramaniam's work on the techniques of iron casting and the effectiveness of canons in the military field. R. Balasubramaniam's article "Massive Forge-Welded Iron Canons of India" has catalogued some of the massive forge welded cannons of the Indian subcontinent. These are marvellous examples of the skills of the Indian blacksmiths in the art of forging large iron objects. The cannons catalogued by Neogi in 1914 have been reviewed. Brief descriptions of the cannons located at Dhaka, Mushidabad, Bijapur, Bidar, Gulbarga, Thanjavur, Bishnupur, Jhansi and Hyderabad have been provided. The studies that need to be undertaken to shed further insight on history, science and technology of cannon technology have been elucidated.

Surinder Singh's article "Scientific Attitudes in Medieval North India: Context, Nature and Articulation (AD 1000–1500)" analyses the nature of scientific attitudes in medieval north India during the pre-Mughal times. This exercise consists of three parts. The first part constitutes a comparison between the observations of Alberuni and Amir Khusro on the levels of scientific attainments pertaining to their respective societies. The second part assesses the contribution of Bhakti reformers in developing a scientific attitude among the lower social strata. The third part explores the role of the ruling classes of the Delhi Sultanate in evolving a scientific outlook towards the twin problems of governance and economy.

S. Jay Kishan's article on "Mining and Metallurgy in Medieval Deccan" sheds light on an important point that is under-researched. In the medieval period the iron industry attained prominence. The application of art and science to this metal for extraction of ores, refinement, alloying, shaping into various articles and applying extraordinary craftsmanship was in vogue in the Deccan in the medieval times among native artisans. The availability of good quality of iron ore in abundance in superficial layers of the crust of the earth and its easy extraction were the main reasons for the development of the iron and steel industry in the Deccan.

The relationship between science and Islam has been the focal point of scholarly exertion and debate for quite some time. S. Irfan Habib's article "Science and Islam: From Pluralism to Religious Essentialism" looks at how science reached its pinnacle with regard to during a period marked by state sponsored free thinking, borrowing from all possible sources available at the time. Thus the early Muslim civilisation, including its science, was heir to a rich and diverse intellectual stock – Roman, Greek, Indian and Persian – accomplishing a unique synthesis of ideas in all branches of knowledge. Several groups and individuals, particularly in South Asia, are questioning this plural legacy today. Some of them are ready to disown their past. Even contributions of early Muslim scientists are being dubbed as works of heretics. Literalism has so profoundly taken over a section of Muslim scholars that they are ready to rewrite their history of science in the light of their skewed understanding of Islam. In his article S. Irfan Habib discusses one such major attempt being made in India through a journal called the *Journal of Islamic Sciences* being published from Aligarh since the mid-1980s.

Dhruv Raina's article "The Needham–Hora Controversy: Disagreements over the Chronology and Approach to the History of Science" is the second in a series of articles on UNESCO, Needham and the first decade of the history of science as a discipline in independent India. This article investigates in detail the first symposium on the history of science in South Asia organised by UNESCO and the National Institute of Sciences in November 1950. It takes up a disagreement on historiography and chronology that cropped up between Needham who reviewed the articles after the symposium and the organisers of the conference. This disagreement in historiography reflects different programmatic visions of the history of science and historical methods. Further, it set the tone for the scientist's history of science that remained the norm within the official history of science in India for several decades. This was quite at odds with the Needhamian historiographical project that was committed to a staunch interdisciplinary one, adopting of humanistic methods in the study of the history of science, and finally the development of a robust historiography if history of science was to matter to science. The controversy died out very fast and none of the issues were resolved. Indian scientists eager to proceed with the task of initiating research in the history of science at that juncture sought the imprimatur from UNESCO and Needham to pursue the history of science of the non-west.

Debasmita Patra, E. Haribabu and P.K. Basu in their joint article titled "Building Interdisciplinary Scientific Specialties: A Case Study of the Solid State and Structural Chemistry Unit (SSCU) IISc, Bangalore" argue that a sociological study of the formation of a scientific specialty and understanding its dynamics and historical sensitivity is very important. Similarly, for a historical study of science, sociological insights are needed in relating science to its context. This article is based on a study of an interdisciplinary specialty and its institutionalisation of cognitive and organisational dimensions in India. It attempts to identify the factors that contribute towards the formation and development of a scientific specialty and its practice in different contexts. To understand the dynamics of solid state chemistry, a research specialty, the article focuses on a department, that is, the Solid State and Structural Chemistry Unit (SSCU), IISc, Bangalore. It also examines how for the first time in India, the SSCU was as a department, exclusively devoted to teaching and research of solid state chemistry was established in 1976 in one of the premier science institutions of India, that is, IISc (Indian Institute of Science). This study takes into consideration twenty-five years of the establishment of the SSCU, that is, from 1976 to 2001, to understand the dynamics of the research agenda of the scientists and their national and international linkages in terms of collaboration.

Arun Bandopadhyay's article "Technology of Defence Production in India under British Rule: Cossipore Factory in the Late Nineteenth and Early Twentieth Centuries" has clearly spelt out the markers that differentiate. An attempt is made in this article to study the technology of defence production in India under the British rule. The article discusses the crucial features of technology of defence production in Cossipore Factory (known as Gun and Shell Factory from 1905 onwards) in the colonial context in India in the late nineteenth and early twentieth centuries. Bandopadhyay has argued that there was a basic dichotomy in the technology of defence production and the general dependence of that production as promoted by colonialism.

Nasir Tybji's article entitled "Jawaharlal Nehru and Science and Technology: The 1958 Scientific Policy Resolution Reconsidered" discusses how the Nehruvian vision of science helped to secularise India. Even though the postulates of the Scientific Policy Resolution themselves required to be restated in accordance with the social aspirations of the time, there is a considerable modification of the original Nehruvian vision and changing audiences over a period of time which prompts us to have a relook. This would then help us to understand the vision and the changing context in view of the long term implications of this policy that constitute a landmark in the history of India.

We are not unaware of the limitations of the discipline with which we are concerned. Unlike the hardcore science, history of science and technology cannot claim for itself the elusive status of objectivity. What we have tried to attempt is a close approximation of our perception of objectivity. Over the last one decade and more, the discipline of history became the worst sufferer of chauvinism, parochialism and sectarian prejudices. We have lost much of our professional credibility. More significantly, the country has witnessed bloodshed because of a highly emotive projection of history. This is the time for atonement. Having made this effort let me admit that, our work in this discipline is rather marginal, more than the intellectual contexts of the pre-modern science. We have initiated a debate on our scientific ethos and technological advances which will hopefully strengthen our pluralistic intellectual tradition. The Sanskrit saying *Vāde Vāde Jāyatç Satyam* (the truth comes out of the debate) has not lost its relevance in the troubled times.

2

Narmada *Homo Erectus*

The Most Primitive Human Fossil of India

Arun Sonakia

There has been a very long history of search for human and pre-human primate fossils throughout the new and old world continents. Whenever a discovery of such a fossil was announced it presumably had a very great impact. It was thought that it would change the views upheld about the human evolution. But it rarely happened. Each and every one of such discoveries did have an impact but it was just proportionate to its significance. Nevertheless there have been a set of criteria and norms that defined and delimited the significance like the location, the chronological context, morphological features to name a few.

Hierarchy of Evolving Humans and Pre-humans

The human and pre-human legacy is sufficiently long. We might recollect that in the early quarter of the preceding century *Ramapithecus* was thought to be the lone beginner of the humans. It lived 14 to 8 million years ago.[1] With time its significance faded and it was realised that it was forerunner of the orangutans that migrated to a place none other than Borneo, Java, Sumatra; the Javan island being one of the locales where the early man originated in South East Asia and presumably migrated and spread to the adjacent continental lands which had partial connectivity with these islands. Small but remarkable studies were made on the molar walls which revealed that they were just vertical and did not bulge bucco-lingually like the humans. Thus the whole genus sunk into the taxon *Sivapithecus*.[2] The taxon *Ramapithecus* had one specific peculiarity that it was widely distributed globally. Its fossils have been found in many countries of the old world continents Africa and Eurasia: Kenya, China, Pakistan, and India to name a few.

The gracile species of genus *Australopithecus* the *Australopithecus africanus* was almost human barring speech. It evolved about 4 million years ago in south and east Africa. Lucy is one famous example of *Australopithecus africanus.* It had bipedal gait but the chin was under developed and unable to generate speech. Then there are reports of subsequently younger antiquity (nearly 2.2–2.0 million years old) named as *Homo habilis* from very localised occurrences of East Africa.

Although the oldest human fossils of the entire world have been reported from Africa which is nearly 1.8 million years year old and the younger from Olduvai George being nearly 1.2 million years old and many more to add after that there appears to be a lull or almost a void as if the whole scenario of human evolution shifted to South East Asia. Although recently there are reports of fairly well preserved human fossils of even 1.4 million years antiquity from Georgia, in South East Asia particularly in Indonesia and China there is a well established continuity of habitation and persistence of early humans, the *Homo erectus.* With a phase of stockily built *Homo sapiens* the Neanderthals appeared around 0.2 to 0.4 million years ago exclusively confined to Europe, the modern population evolved with the advent of the *Homo sapiens* 0.2 to 0.3 million years ago in Africa, Asia, and Europe.[3] There are many theories of dissemination and spread of the *Homo sapiens* such as the "Out of Africa" theory as per which the humans continued to migrate from Africa to other countries in pulses.

History of Search

The search for the skeletal remains of early humans commenced at the instance of Charles Darwin way back in 1858. As per Darwin[4] the humans, the apes and monkeys had a common ancestor and the "missing link" that connected us with those ancestors who lay buried somewhere in the warmer latitudes of central Africa. Eugene Dubois, a young Dutch doctor was so impressed by this hypothesis that he joined the American Army and reached Java in about the same middle latitudes. In 1892 he discovered a femur and a part of a skull confirming the existence of the missing link between humans and pre-humans as quoted in Hardlicka.[5] Then there was a row of discoveries on the island. It was in late twenties that reports of discoveries of human fossils started coming in from mainland China popularly known as the "Peking man". In the late fifties reports of discovery of early human fossils started coming in simultaneously from south-central Europe and East Africa. Nearly a decade before the pioneering discovery of *Pithecanthropus erectus* nee the Jawa man we lost the chance of such a great discovery from India. Theobald[6] mentions "a human cranium was lost from the museum of Asiatic Society of Bengal before it could be described." A century later a human fossil was discovered by the author from the Middle Pleistocene gravels belonging to the Surajkund formation exposed in the river channel near village Hathnora of Sehore district in MP. This place is in fact 80 kms southeast of Bhopal.[7] There are reports of a much younger human mandible from Magasaryanigondi cave deposits in Kurnool district of AP.[8] Then we have Holocene burial sites of Pratapgarh, Dholavira and the Harappan antiquity. Still later there are reports of agricultural village cemeteries of Inamgaon and others.

Significance of the Indian Human Fossil from Narmada Valley

The significance of Narmada *Homo erectus* fossil could be assessed on two important counts: one, the time and the other space. Discovered almost midway between the thousands of kilometres apart human fossil occurrences of the Lavent zone of West Asian countries like Israel on one side and on the east South-east Asian countries of Indonesia, China on the other side of the globe, the Indian fossil is best suited to provide clue to the evolution, migration and, dissemination of the human species. Till date this is the only authentic record of the human fossils in the subcontinent and above all it is the most primitive human fossil of India and the adjacent countries and no other immediate predecessor or the successor have been discovered so far.

Comparative Morphological Studies

The Narmada human fossil comprises the most vital component of the skeleton, the skull. It is fairly complete to outline and define conclusively its outer contours to reconstruct its shape and size and also to assign it to a specific taxon. Being highly mineralised it has turned grayish brown in colour The cranial bones are very thick at parietal margin, 13 mm. It is three times thicker than that of the modern humans. The feature that is very well developed is the forwardly projecting eyebrow ridges. The Glabella is broken and a large void of frontal sinus is exposed although filled with some glossy smooth material. The cranial roof is in fact supported by internal accumulation of the sediments in which it lay buried The frontal is gently sloping and on the coronal section the parietals as also the temporals are falling apart thus making scope for a wider base The broader base is a morphological character typical of *Homo erectus* as with evolution the base of skull contracted and the upper part grew bulbous to accommodate a larger brain in the coming generation of *Homo sapiens.* The temporal lines are very strong culminating posteriorly in to a very strong *torus angularis*. The mastoid like the primitive crania are underdeveloped. Ear holes and the eye sockets are rather large. The skull as a whole is long and low but the angularities at the sagittal suture that itself lies in a deep furrow and where the parietal squama turns downward, are very prominent. A deep fissure lies between the petrosal crest and the mastoid process. In the occipital region most prominent zone that is generally situated very low is placed very high at the level of opisthocranion. Though abraded the occipital torus is prominently developed.[9]

Comparison with Other Crania and Present Status

The morphological features and the indices of metrical dimension of the Narmada skull compare very well with Asian *Homo erectus* crania from Indonesia and China. In certain characters it shows little advanced stage of development in comparison to the Sangiran 17 from Indonesia skull but in many features and indices it has similarities with the *Sinanthropus peninensis.* Comparative studies have been carried out with the casts of African *Homo erectus* KNM ER 3733, the European specimen from Arago with due assistance of the French palaeoanthropologist De Lumley. The author finds enough scope to differ with American palaeoanthropologist Kennedy[10] who would put it as Archaic *Homo sapiens* based on multivariate analysis of data.

Age and Antiquity of the Narmada Hominid

The Narmada human fossil has been dated on the basis of mainly two criteria. One is the associated fauna and the other is palaeomagnetism. Amongst the big list of Quaternary mammalian fauna that would include mainly the cattle there are species of *Bos namadicus, Bubalus, antelopes Cervids, Pigs, Proboscidians* the *Sus namadicus,* and the *Stegodon namadicus* are typically index fossils of Middle Pleistocene age.

The other parameter of age determination is palaeomagnetism. The Goss-Matuyama palaeomagnetic reversal has been dated at 0.76 million years and the Dhansi formation where this reversal has been marked lies just below the basal gravel bed that yielded the human cranium.[11]

Notes

1. G.E. Pilgrim, 'Notices of New Mammalian Genera and Species from the Tertiaries of India', *Records of the Geological Survey of India* 40, 1910, pp. 63–71.
2. D. Pilbeam, G.E. Meager, C. Bagley, M.D. Rose, M.H. Pickford, A.K. Behrenmeyer and I.S.M. Shah, 'New Hominid Primate from Siwaliks of Pakistan and their Bearing on Hominid Evolution', *Nature* 270, 1977, pp. 689–95.

3. Susan C. Anton, 'Natural History of Homo Erectus', *Yearbook of Physical Anthropology* 46, 2003, pp. 126–270.
4. Charles Darwin, *On the Origin of Species*. London: John Murray, 1859.
5. Ales Hrdlicka, *The Skeletal Remains of Early Man*. Washington, DC: Smithsonian Institution, 1930.
6. William Theobald, 'The Siwalik Group of Sub-Himalayan Region', *Memoirs of the Geological Survey of India* 14, 1881, pp. 66–125.
7. Arun Sonakia, 'The Skullcap of Early Man and Associated Mammalian Fauna from Narmada Valley Alluvium, Hoshangabad Area M.P.', *Records of the Geological Survey of India* 113(6), 1984, pp. 159–72.
8. Anonymous, 'News', *Geological Survey of India, Southern Region* 123, 1994, p. 9.
9. M.A. De Lumley and Arun Sonakia, 'Premiere decouverte D'un Homo erectus sur le continent Indien a Hathnora dan la Moyenne Vallee de la Narmada', *L' Anthropologie* 89, 1985, pp. 13–61.
10. Kenneth A.R. Kennedy, 'Is Narmada Hominid an Indian Homo Erectus?' *American Journal of Physical Anthropology* 86, 1991, pp. 475–96.
11. Arun Sonakia and S. Biswas, 'Antiquity of the Narmada Homo Erectus: The Early Man of India', *Current Science* 75, 1998, pp. 391–93.

3 Megalithic High Tin Bronzes

Sharada Srinivasan

It is generally understood that there was a hiatus and even regression in terms of developments of copper-bronze metallurgy between the Indus Valley (3rd millennium BC) and the later archaeological contexts in the Indian subcontinent of the subsequent chalcolithic and megalithic cultures of the second and first millennium BC in terms of range or quantity of finds. Even so, this article puts together investigations on some bronze vessels from Iron Age Indian sites and megalithic contexts of southern India such as the Nilgiri megaliths and cairns and the Adichanallur Iron Age burial sites to show that these reflect sophisticated bronze working practices with the use of a specialised alloy known as wrought and quenched high-tin bronze alloy, i.e., (hot forged and quenched binary copper-tin alloys of about 23% tin resulting in the predominant retention of martensitic beta phase). These rank not only amongst the earliest such alloys known in the world but also suggest that bronze metallurgy at the time was more advanced than previously thought.

Previously it had been thought that the skilled technology of high-tin beta bronze working was developed in South East Asia or China and then spread to India, while its use is scarcely reported outside Asia. However studies by the author of surviving techniques of making wrought and quenched beta bronze bowls and high-tin delta bronze mirrors in parts of Kerala together with analyses of finds from South Indian megaliths and Iron Age burial sites from other parts of India from early to late first millennium BC suggest that the technique might have had indigenous origins (Srinivasan, 1994, 1997a, 1998a; Srinivasan and Glover, 1995, 1997). Vessels from Adichanallur and Nilgiris represent some of the finest examples of wrought and forged high-tin bronze vessels found anywhere with very fine rims of down to 0.2 mm thick indicating extensive forging followed by quenched and fine decorations. If the above were indeed so, it also remains a possibility that this technique could have in turn also spread to tin-rich southeast Asia, especially Thailand where well documented finds of high-tin bronzes are also known and where the technique seems to have flourished to skilled heights in its own right at Ban Don Ta Phet by the 4th century BC (Rajpitak and Seeley, 1979; Bennett and Glover, 1992).

Investigations on Megalithic High-tin Bronzes

Generally as-cast binary copper-tin alloys with over 15 per cent were not much in vogue in antiquity as they became brittle due to the presence of the delta phase component. However investigations by the author from pre-historic, medieval to modern South India indicate the continued use of specialised binary high-tin beta bronzes with 22–25 per cent tin to make artifacts such as vessels, coins and musical instruments right into the present day (Srinivasan, 1994, 1998a).

Metallurgical investigations by the author on very thin vessels of a thickness of 0.2–1 mm from South Indian burials and megaliths of Adichanallur and Nilgiris of the early to mid first millennium BC and medieval Chola platters (10th–12th centuries) (Srinivasan, 1994; Srinivasan and Glover, 1995) indicated that these were wrought and quenched high-tin beta bronzes, i.e., copper-tin alloys with 23–25 per cent tin. These were fabricated by extensively hammering out such an alloy between 586–798°C when plastic beta intermetallic compound (Cu_5Sn) of equilibrium composition (22.9% tin) forms, followed by quenching resulting in the retention of needle-like beta phase which prevents the formation of brittle delta phase. In contrast low-tin bronzes have limited workability.

Despite the analyses reported in Breeks (1873: 63) of a few vessels from the Nilgiri cairns being of 20–25 per cent tin-bronze over a 100 years ago, it was believed that these artifacts were imported (Leshnik, 1974: 156) and the possibilities of there being an local continuing tradition had never been explored. The author was the first to have identified such a continuing tradition of high-tin beta bronze vessel making from any part of the world, in the village of Payangadi in Kerala (first reported in 1991: Srinivasan, 1998a; this activity has now sadly ceased to take place at this village) and to have correlated these with micro-structures in vessels from the South Indian megaliths of Adichanallur and Nilgiris (Srinivasan, 1994; Srinivasan and Glover, 1995). Thus, the vessel was made by extensively hammering out, in cycles of annealing and hot forging, an alloy close to a composition of 23 per cent Celsius around 600°C in the temperature range of formation of the plastic beta intermetallic compound, followed by quenching, resulting in the retention of needle-like beta phase. At the workshop the author also found that the resultant alloy has musical properties and takes a very bright golden polish which would explain its significance in antiquity.

Although prior to these studies non-Indian sources had been suggested for the bowls from the Nilgiri megaliths in West Asia (Leshnik, 1974: 156) or East Asia (Rajpitak and Seeley, 1979) it now seems most likely that they were made in ancient Kerala since they closely resemble the wrought and quenched high-tin beta bronze bowls from Payanagadi as observed by the author, not just metallurgically but also stylistically in the use of concentric rings in the centre of the bowls made in 1991 at the workshop using a hand-turned lathe. This is consistent with local lore where the Todas, the original inhabitants of the Nilgiris, claimed that such vessels came from Kerala (Srinivasan, 1994); the author is currently investigating vessels from megaliths in Kerala. Allchin and Allchin (1982) also postulate indigenous developments for the South Indian megaliths.

As for better dated sites, the author is currently investigating a few vessel fragments and artefacts at the Vidharbha megaliths of Maharashtra in the northern Deccan, with a single or two carbon dated to the 8th Century BC; the micro-structure of a finial from Boregaon shows it to be a cast and quenched beta bronze with 21 per cent tin, while a couple of wrought and quenched beta bronze bowls also identified by the author from Mahurjhari with 18–19 per cent tin show some martensitic quenched beta phase amidst alpha phase islands. However in none of these high-tin bronze vessels from the Vidarbha megaliths was the composition range of formation of predominant beta phase of around 23 per cent tin fully reached nor were they as much worked or hot forged as the high-tin bronze vessels from the Iron Age or megaliths from Tamil Nadu or Nilgiris and Adichanallur which show a stage of true mastery of the plasticity of the martensitic beta phase. Two samples excavated from the Indus Valley site of Mohenjodaro (*c*. 2500 BC) were also analysed and found to be 22 per cent tin-bronze (Mackay, 1938: 480), although micro-structural data is not available. These finds may predate those hitherto known from elsewhere in antiquity. This may suggest a departure from well entrenched ideas (e.g., Craddock et al., 1989: 23) that the Indian subcontinent would have more likely received diffusionist bronze technologies or traded bronze from tin-rich Asia, especially South East Asia.

Links with Present Day High-Tin Beta Bronze Bowl Making in Kerala

In 1991 the author documented the making of wrought and quenched high-tin beta bronze vessels, gongs, cymbals and ladles in Payangadi in Kerala. Microstructural investigations on two bowls from Payangadi and Trichur confirmed that the largest concave bowls, of 25 cm, 8 cm deep and 1.4 mm rim thickness, were made out of a flat circular ingot of 22.5 per cent tin of a diameter of only 15 cm, and 1.5 cm thick, by forging in cycles of hammering and annealing followed by quenching between 600–700°C, all in the $\alpha+\beta$ phase field. The finished bowl indicates the remarkable degree of plastic deformation which was achieved to more than twice the dimensions of the ingot, the size of which is demarcated in the base of the concave bowl. The micro-structure of the vessel confirmed that it was of wrought and quenched high-tin bronze of about 23 per cent tin, with beta phase needles. In a collaborative study done by the author with Oleg Sherby it was found that high-tin bronze is quasi-superplastic.

The inner surface of quenched bowls are often polished bright golden, to contrast with the exterior quenched skin. The bright golden lustre on polishing would explain why this alloy was in vogue in ancient times whilst it is remarkable that in one Iron Age example from Adichanallur and a couple from the Nilgiri megaliths, the dark patination rich in tin oxides which are often found, is not yet formed and the original bright golden lustre still comes through as seen in the Nilgiri bowl.

My late grandmother, a householder interviewed in 1991 said that the *talavettu* (i.e., musical vessels due to the tonality of the β-martensitic alloy) were preferred to brass for storing food, very likely due to the non-toxicity of a high tin content and the alloy's corrosion resistance of the alloy. She added that food was not heated in such vessels, probably due to formation of embrittling lower temperature phase upon cooling. Moreover, at the workshop in Payangadi visited in 1991, the inner part of the wrought bowls were decorated using a hand turned lathe mounted on a wall and inscribed with concentric rings in a manner very similar to the megalithic bowls from Nilgiris and Adichanallur. Thus it was possible to extrapolate how the South Indian megalithic high-tin bronzes could represent a longstanding local metallurgical tradition.

In 1998 when the author and Ian Glover re-visited Kerala, these craftsmen from Payangadi had given up their trade. However, another set of craftsmen were at work making rather smaller high-tin bronze vessels in Paridur in Kerala. Here to speed up the process a set of four ingots were piled on one another and forged and then the final forging was done on each before quenching.

Re-examining the Milieu of Megalithic High-tin Bronzes

The emergence of urbanism in southern India and the factors leading to it are perhaps less well understood. The process of urbanisation from neolithic or megalithic stages seems to have come about later by about the 3rd century BC with relatively few habitation sites having been found close to megalithic sites. However, it is interesting that even in the absence of typical urban characteristics such as structural remains, there is some evidence for other features associated with urbanism such as writing, literature and evidence of long-distance exchange. For example, forms of graffiti or writing thought to be related to Tamil Brahmi are claimed to be found on pot sherds from sites such as Kodumanal. Intriguingly, of writing resembling Tamil Brahmi were reported to be found from three urns excavated in 2004 by T. Satyamurthy, ASI, Chennai, from an impressive urn burial site of Adichanallur that may go back to the 7th or 8th century: which may throw new light on the origins of writing in South Asia according to S. Deraniyagala of ASI, Sri Lanka (Subramanian, 2004). Such findings may also have significance with respect to understanding the emergence of the fine but enigmatic corpus of Tamil Sangam literature, loosely dated to the 5th century BCE to 5th century CE. That the early Sangam era may have coincided with the iron age megalithic period (first millennium BC) is suggested by references to stone megaliths: including urn-burials and cyst-burials (Narasimhaiah, 1980: 188).

Findings from megaliths in southern India point to long-distance exchange networks even in an era slightly preceding the true early historic period. Berenike in Egypt principally traded in stone and quartz beads from sites in Tamil Nadu such as the megalithic burial-cum habitation site of Kodumanal (*c.* 300 BCE) and in glass beads from Arikamedu until the 2nd century (Francis, 2002). Carnelian and lapis lazuli found from megalithic Kodumanal probably had their sources in western India and Afghanistan respectively. A copper alloy tiger figurine from megalithic Kodumanal inlayed with carnelian and lapis lazuli may be seen to share affinities with an early historic carnelian tiger found in Thailand and reported in Glover (1990).

The high sophistication of numerous vessels and gold jewellery uncovered from the Nilgiri cairns (loosely dated from about 1000–500 BCE) and Adichanallur burials (*c.* 1000–800 BCE) had led scholars to speculate whether they were imported (Leshnik, 1974; Knox, 1985; Rajpatik and Seeley, 1979), especially since the best documented studies on high-tin bronzes has come from Ban Don Ta Phet in Thailand, *c.* 4th century BC. However this author has argued (Srinivasan and Glover, 1995; Srinivasan, 1997), as also indicated in this article, that technical studies now suggest that it is most likely that the analysed vessels, which were of high-tin beta bronze, were of local provenance which is also supported by evidence of a longstanding local tradition continuing into the present day, especially in Kerala. In fact, the Nilgiri and Adichanallur vessels rank amongst the most elegant and early examples of high-tin bronze metalware. The collection of vessels from the Nilgiri hoards in the British Museum numbers no less than 40 and in particular comprise elegant fluted, knob-based and carinated vessels with concentric rings etc., while the elongation and degree of hot forging in these is far greater than seen in the high-tin bronzes from Thailand suggesting a greater mastery of the technique in the South Indian megalithic high-tin bronzes. At any rate, evidence for long-distance contact with South East Asia is suggested in the stylistic similarities that knob-based and ringed bowls from the Nilgiri cairns showed with vessels from Ban Don Ta Phet (*c.* 4th century BCE). This suggests that maritime interactions between southern India and South East Asia were prevalent even in an era pre-dating the better documented Roman maritime trade.

Ray (1996) has suggested that rather than attributing social change in peninsular India primarily to Mauryan intervention, the impetus from coastal and maritime interactions should also be taken into account. A major trade route in the early historic period was the route from the Malabar coast in the west through the Palghat Pass, and moving east to the Kaveri delta with its settlements and including sites such as Kodumanal, Karur, Arikamedu, Korkai and Alagankulam (Thapar, 2002: 237). Parasher–Sen (2000) also point to the significance of local developments in the Deccan in terms of origin of settlements. Tamil Sangam works such as the Ettuttokai refer to a prosperous people living in Tamil Nadu and Kerala who not only had seafaring trade with the Yavanas or Mediterranean people but also with the island of Savakam, thought to be Java in South East Asia (Marr, 1985; Nagaswamy, 1995b: 96). The above discussion also suggests that the existence of skilled metallurgical or craft techniques need not necessarily be linked with settled urbanism and indeed, even till recently in India many crafts such as metal crafts were practised as itinerant or migrant trade.

Postulating a Musical Significance for the Megalithic High-tin Bronze Vessels

While the megaliths, not to mention the megalithic high-tin bronze bowls themselves, remain an enigma, another matter for speculation is that of the function of these highly specialised bronze bowls. One possibility could have been ritual function and indeed the notion of the raised knob-base surrounded by rings and the floral lotus patterns seen on some Nilgiri bowls and links with bowls from Thailand, have made some commentators such as I. Glover speculate about ritual associations with Buddhism. This is seen from the decorations typically of the raised knob-base and rings representing mount Meru in the centre. Of course, in practical terms the functionality of high-tin bronze in its use of storing food due to its less corrosive properties has been observed from the studies by the author on current high-tin bronze working practices in Kerala.

However, a third possibility postulated by the author is that the vessels may have been used for their musical significance even as far back as the megalithic period since the quenched high-tin bronze alloy is found to have good musical properties and is used in making cymbals even today as documented by the author with I. Glover in Kerala in 1998. The grounds for suggesting that its musical properties may have appealed even in megalithic times comes from recent publications by Boivin (2004) that clusters of rocks with depressions from the Neolithic site of Kupgal in Karnataka in southern India could have been used for making ringing or percussion sounds.

References

Bennett, A. and I.C. Glover. (1992). "Decorated High-Tin Bronzes from Thailand's Prehistory", in I.C. Glover (ed.), *Southeast Asian Archaeology*. Hull University: Centre for Southeast Asian Studies, 187–208.

Breeks, J.W. (1873). *An Account of the Primitive Tribes and Monuments of the Nilagiris*. London: India Museum.

Boivin, N. (2004). "Rock Art and Rock Music: Petroglyphs of the South Indian Neolithic", *Antiquity* 78 (299): 38–53.

Craddock, P.T., I.C. Freestone, I.C. Gurjar, L. Willlies and A. Middleton. (1989). "The Production of Lead, Silver and Zinc in Early India", in A. Hauptmann, E. Pernicka and A.G. Wagner (eds.), *Old World Archaeometallurgy*. Proceedings of the International Symposium 'Old World Archaeometallurgy', Heidelberg 1987. Bochum: Selbstverlag des Deutschen Bergbau-Museums, 51–69.

Francis, P. (2002). *Asia's Maritime Bead Trade from ca 300 BC to the Present*. Honolulu: University of Hawaii Press.

Knox, R. (1985). "Jewellery from the Nilgiri Hills: A Model of Diversity", in J. Schotsmans and M. Taddei (eds.), *South Asian Archaeology 1983*. Instito Universitario Orientale, Dipartimento di Studi Asiatici, Series Minor XXIII, Naples: Instituto Universitario Orientale, 523–33.

Kuppuram, G. (1986). *Imperial Cholas of Karnataka*. Delhi: Sundeep Prakashan.

Leshnik, L. (1974). *South Indian 'Megalithic' Burials: The Pandukal Complex*. Weisbaden: Franz Steiner.

Mackay, E.J. (1938). *Further Excavations at Mohenjodaro*. New Delhi: Government of India Press.

Malroney, C. (1975). "Archaeology of South India: Accomplishments and Prospects", in Burton Stein (ed.), *Essays in South India*. Hawaii: The University Press of Hawaii.

Mukherjee, M. (1987). *Metal Craftsmen of India*. Calcutta: Anthropological Society of India.

Narasimhaiah, B. (1980). *Neolithic and Megalithic Cultures in Tamil Nadu*. New Delhi: Sundeep Prakashan.

Parashar-Sen, A. (2000). "Origin of Settlements, Culture and Civilization in the Deccan", in H.K. Gupta, A. Parashar-Sen and D. Balasubramanian (eds.), *Deccan Heritage*. Hyderabad: Indian National Science Academy and Universities Press, 233–58.

Paramasivan, S. (1941). "Investigations on Ancient Indian Metallurgy: A Pre-Historic Bronze Bowl", Proceedings of the Indian Academy of Sciences 13(2), Sec. A, 87–90.

Ray, H.P. (1996). "Maritime Archaeology of the Indian Ocean: An Overview", in H.P. Ray and J.F. Salles (eds.), *Tradition and Archaeology: Early Maritime Contacts in the Indian Ocean*. New Delhi: Manohar, 1–10.

Rajpitak, W. and N.J. Seeley. (1979). "The Bronze Bowls of Ban Don Ta Phet, Thailand: An Enigma of Prehistoric Metallurgy", *World Archaeology* 11 (1): 26–31.

Srinivasan, S. (1994). "High-Tin Bronze Bowl Making from Kerala, South India and its Archaeological Implications", in A. Parpola and P. Koskikallio (eds.), *South Asian Archaeology 1993*. Annales Acadaemia Scientiarum Fennicae, Series B, Vol. 271, Helsinki: Suomalainen Tiedeakatemia, 695–705.

——. (1997). "Present and Past of Southern Indian Crafts for Making Mirrors, Lamps, Bells, Vessels, Cymbals and Gongs: Links with Prehistoric High-Tin Bronzes from Mohenjodaro, Taxila, South Indian Megaliths and Later Finds", *Journal of the Society for South Asian Studies* 13: 209–25.

——. (1997). "The Composition of Bronze Slags from Kalyadi in South India, and the Implications for the Problem of Tin in South Indian Antiquity", in A. Sinclair, E. Slater and J. Gowlett (eds.), *Archaeological Sciences 1995.* Proceedings of a conference on the application of scientific techniques to the study of archaeology, Liverpool, July 1995. *Oxbow Monograph Series* 64, Oxford: Oxbow Books, 136–42.

Srinivasan, S. and I. Glover. (1997). "Archaeometallurgical Implications of New Findings of Traditional Crafts of Making High-Tin 'delta' Bronze Mirrors and 'beta' Bronze Vessels in Kerala State of South India", in P. Vandivar, J. Druzik, J. Merkel and J. Stewart (eds.), *Materials Issues in Art and Archaeology V.* Materials Research Society Symposium Proceedings 462, Warrendale: Materials Research Society, 81–105.

——. (1995). "Wrought and Quenched, and Cast High-Tin Bronzes from Kerala State, Southern India, Part I, Contemporary Manufacture", *Journal of Historical Metallurgy* 29(2), London: Institute of Metals, 69–75.

Subramanian, T.S. (2004). "Skeletons, Script Found at Ancient Burial Site in Tamil Nadu", *The Hindu*, May 26, 12.

Swarnakamal, K. (1980). *Studies in Metallic Art and Technology of Gujarat.* Baroda: Museum and Picture Gallery.

Thapar, R. (2002). *The Penguin History of Early India: From the Origins to AD 1300.* New Delhi: Penguin Books.

4 Crafts on the Move?

The Case of the Megaliths of Peninsular India

Jaya Menon

The subject of this article is the yet enigmatic Iron Age culture of peninsular India, known as the megalithic culture, best known through its interment of the dead. Not surprisingly, the megaliths have captured the attention of laymen, explorers and scholars alike since the mid-nineteenth century. Much has been written on the typology of megalithic burials, the distribution of various types, the origin of the port-holed cist, the contents of megalithic burials, the ubiquity of iron and black-and-red ware, the issue of language movements, the paucity of habitation sites, whether these were the remains of sedentary or nomadic groups, as well as whether the term megalith can be used for burials in several contexts such as those in urns. The intention is not to go into most of these issues, relevant as they may be. In this article, I intend to focus specifically on technology, in the face of the prevailing debate about these societies as having had sedentary or nomadic lifestyles.

The megaliths by no means are an easy topic for discussion. The culture, if it can with any justification be called so, is hardly a homogeneous entity spanning a great length of time and encompassing almost as many variations in its embodiments as sites. Chronology is not a settled issue as far as the megaliths are concerned nor is their origin and dispersal, if any. Neither are megaliths restricted to peninsular India as scattered sites are found in central, northern and north-western India. Most of the research has tended to focus on megalithic complexes within modern-day state boundaries such as Andhra Pradesh, Tamil Nadu, Kerala or Karnataka or in some cases, from a region such as Vidarbha and often, early research focused on sites in smaller administrative boundaries such as districts.

No easy demarcation of megalithic types with regions is possible as is clear from several excavations, as sites appeared to have multiple types. Early correlations made between megaliths and iron and black-and-red ware are being re-investigated as it is now realised that both iron and black-and-red ware can in be found from non-megalithic contexts. In fact, Subbarao[1] had pointed out as early as 1957 that the black-and-red ware needed to be seen as separated from the megalithic burials. Actually what should be more important than the mere fact of black-and-red ware found along with megalithic burials is the typology of this particular category of black-and-red ware. Concordance of megalithic burials with black-and-red ware horizons in habitation mounds, then, is possible only when diagnostic types, such as, funnel-shaped lids or the so-called 'tulip-shaped vase' are found in both contexts.

Scholars have also pointed out the intrusive nature of megalithic layers in habitation sites. At Brahmagiri, a major megalithic habitation-cum-burial site, Wheeler[2] found overlaps between not only the 'Stone Axe Culture' and the megalithic horizon but also between the latter and the Andhra Culture. Despite the overlap between the first two cultures, he showed that there was no cultural transition between them. "The sudden and overwhelming character of the megalithic intrusion upon the Stone Age natives of Brahmagiri is abundantly manifest"; in fact, he considers the megalithic as an invasion. This kind of overlap is noticed at other sites too, such as at Kesarapalle[3] but again distinctive cultural traditions are connected with the arrival of a new people or at Nagarjunakonda where again there is no sequential break in stratigraphy of the burials (there is no habitation mound) but pottery types and iron implements are illustrative of a break.[4] At Maski, Thapar[5] points out the sudden emergence of megalith-building people and the introduction of iron and notes that the maturity of the cultural material suggests the intrusion of a perfected tradition.[6] In this context, it is interesting to remember that an early Iron Age is generally characterised by few and small hunting and fishing tools.[7] However, iron is clearly typologically advanced in megalithic burials and yet this period sees the introduction of iron in peninsular India. In that sense, it is somewhat tempting to consider such a "perfected tradition" as originating from elsewhere.

The chronology of the megalithic complexes is of some relevance and needs to be briefly recapitulated. A relative chronology was constructed first at the habitation-cum-burial site of Brahmagiri by Wheeler who, on the basis of stratigraphy and the find of coins in his Andhra layers, attributed a date range of the second century BC to the mid-first century AD for the megalithic occupation. This was largely adhered to by several subsequent researchers till excavation at the site of Hallur gave dates of about c. 1000 BC for the black-and-red ware horizon. According to Ghosh[8], most C14 dates for the megalith can be placed between 1000 BC (Hallur) and AD 1000 (Pykara), but the period of "maximum popularity" is considered as 700 years from 600 BC to AD 100. However, in South India, there is clear overlapping between the megalithic and the early Historic and in Tamil Nadu and Kerala, the Early Historic itself largely overlaps the period of the Tamil Sangam texts dated to around the first to third centuries AD. It would appear that the upper end of dates for megaliths can largely conform to Wheeler's at Brahmagiri while the lower end needs to be revised and may go back to about 1000 BC. This is of particular relevance for this article, as the last phases of the megalithic culture appear to coincide with the period of Sangam literature[9] attesting to a high level of socio-economic organisation.

The Agriculturalist/Nomad Debate

The argument for an agricultural base for the economy of the megalithic culture has rested largely on the distribution of certain megalithic sites as well as on the evidence of crops among botanical remains. Certain sites were pointed out as located on hilly slopes or rocky outcrops on high ground, usually in association with large irrigation tanks.[10] There were two implications here: one, that the burials occupying rocky high ground meant that the few stretches of plain open land would have been left for agricultural purposes indicating a concern of the megalith builders with conserving cultivable land. The second implication was that the location of burials near irrigation tanks meant, *ipso facto,* the use by the megalith builders of the tanks themselves. Banerjee[11] points out that "it may not be wrong to regard them (the megalith builders) as the introducers of the irrigation system in the south by the tank irrigation method, which involved the collection of rain-water flowing down from the surrounding catchment-area by bunding up the sloping side." According to Srinivasan and Banerjee[12], the larger the irrigation tank, the larger the cluster of megaliths nearby, implying a direct correlation between megaliths and tanks. Rao[13] has also laid out some of the evidence, testifying to the agricultural basis of the megalith builders. These range from agricultural tools found in megalithic burials, to the etymology of place-names of megalithic sites, to time and labour issues involved in the construction of megalithic monuments. It also points to the lack of discernible evidence as to what megalithic folk would have offered in exchange if they were procuring pottery and other materials as well as grains from settled peoples. A final argument was also that several crafted objects found in megalithic burials implicitly meant a sedentary existence.

The relationship of tank irrigation with megaliths has, by no means, been universally agreed upon. For one thing, tank irrigation is limited to certain parts of South India and cannot be considered universal. Moreover, instances have been found[14] of sites that are apparently on the edges of tanks in the summer season but would be submerged in water during the rainy season. Similarly, those sites in the foothills near valleys would have been submerged if bunds were constructed across the valleys.[15] It also appears an exaggeration to say that all megalithic sites occupied only rocky upland regions, as examples have been given[16] of burial sites in open fertile areas as well.

The main protagonist for a pastoral background for the megalithic complex is L.S. Leshnik. For him,[17] the evidence rests on the rarity of habitation sites in contrast to the cemeteries. There are problems about directly correlating black-and-red ware horizons at habitation sites with megalithic burials (or in other words of assuming that black and red ware would be found only with megaliths); the predominance of weapons or tools of offence among the interred iron objects, the rarity of agricultural tools, as also the fact that tools like bill-hooks can be mainly attributed to pastoral peoples or those living in forested regions rather than with people practising extensive plough agriculture.

It does appear that not enough concerted attempts have been made to study technologies in the Iron Age in South Asia. Most reports on megalithic sites focus on typologies of crafted objects, leaving aspects of technology less well delineated. The assumption is also often made that the very fact that crafted objects were found within megalithic burials meant those crafts were practised, which in turn implied necessarily a settled existence. This article is a preliminary effort to examine this assumption of craft production and sedentary living with the initial disclaimer that no solutions or answers will be provided. Instead, perhaps, many more questions will in reality emerge.

Leaving aside the issue 'that megalithic complexes in peninsular India can hardly be considered a homogeneous category, we note that the major crafts appear to be pottery production, metallurgy' metal working (of copper and iron) and bead-making. While there are crafted objects found in other materials (such as shell, glass, gold and silver), there is as yet little evidence of their production in the form of debitage, or unfinished objects or raw material stocks or facilities.

The Archaeological Evidence

The agriculturalist argument for megalithic complexes, as pointed out, rests primarily on the habitation sites that have been discovered. Most habitation sites have multiple periods of occupation with the black-and-red ware horizon in many cases not insubstantial. At Brahmagiri, for example, Period II "considered identical with the local megalithic tombs and pit circles"[18] was about 0.9–1.2 m in thickness.[19] With the usual disclaimers about simplistically correlating depth of occupation with time duration, Wheeler tentatively suggests a duration of about 200 years for Period II. But as correctly pointed out by Sundara[20] and evident from Wheeler's[21] own analytic table of frequencies of sherds of the three different cultures there hardly seems to be any independent megalithic horizon as the lower levels of the black-and-red ware layers overlap with Period I and the upper layers with Period III.

The list of habitation sites and habitation-cum-burial sites from Moorti[22] was useful for the purpose of craft analyses, even if some sites could not be categorised easily. Out of a total of possibly 43 habitation-cum-burial sites, it was recorded that 38 gave some evidence of iron smelting and working, 2 for copper working and smelting and 5 for gold. Out of a total of 12 habitation sites, 8 gave evidence for iron smelting and/or working, 1 for copper, 2 for gold and 1 for silver.[23] The evidence for gold was largely in the form of 'ancient gold workings', an unspecified category.

Moorti[24] points out elsewhere that a large number of habitation and habitation-cum-burial sites were located within a distance of 0–20 km from iron ore sources. At only a few sites do we get concrete evidence for craft – at Naikund, for example, an iron smelting furnace was found, at Peddabankur – Kadambapur, a blacksmith's workshop, as also at Bettada Banahalli. Kodumanal gave much evidence for smelting in the form of furnaces, slag, and tuyeres. Vitrified crucibles may give evidence for the production of steel.[25] At most other sites, remains are in the form of slag that provide some of the best evidence for iron smelting.

For pottery evidence, there is the evidence of Polakonda and Bettada Banahalli but at both the craft evidence is apparent from neolithic levels and probably continues into the megalithic occupation.[26] At Kesarapalle, in Krishna district, Andhra Pradesh, a stone dabber was found from the surface (which, admittedly, may not be megalithic) and a so-called 'colour-grinder' from the megalithic layers.[27] From Sanur[28] a similar object termed as a circular hollowed quern of granite from a megalith was found. However, the objects appear to be stones with depressions into which the pivot of a wheel would fit. Both the sites have reported wheel-turned pottery from megalithic layers.

As far as bead-making is concerned, the two sites that are normally considered as centres for the craft are Mahurjhari in Vidarbha and Kodumanal in Tamil Nadu. Both are habitation-cum-burial sites though in the case of Mahurjhari the mound has not been excavated and the concordance of any layers with the burials is unclear. Yet the habitation has been adjudged as clearly Early Historic. Kodumanal's habitation layers have been correlated with the burial evidence. At Mahurjhari, Deo[29] found that typologically beads seemed to belong to the Early Historic period, but because a number of unperforated etched and unetched carnelian beads were found in the megalithic field, he presumed that beads were made in the Iron Age too. Subsequent work done at Mahurjhari[30] recovered a bead manufacturing locality some 500 m away from the Early Historic habitation mound, where ample evidence of manufacture in the form of finished, unfinished artefacts, semi-processed raw material and debitage was found spread over an area of 90 × 50 m. Mahurjhari had access to the plentiful crypto-crystalline silicious stones that are a part of the Deccan Trap. Kodumanal in Tamil Nadu had access to important quartz bearing and beryl bearing sources in its vicinity. Carnelian, however, was not locally available and was probably brought from outside. Possibly the numerous etched carnelian beads found at Kodumanal were manufactured locally, as there is some small evidence of carnelian working, though there is far more evidence for production from the locally available quartz. Interestingly, though quartz debitage has been plentifully found in the habitation mound, there is no such evidence from the megalithic burials at Kodumanal.[31] Thus, the evidence for megalithic bead-making is limited.

There is some evidence for other crafts such as spinning and weaving. At several sites circular (annular) terracotta rings with perforations too large for them to have been beads have been found. These have been found from Brahmagiri[32], Kesarapalle[33] and Nagarjunakonda.[34] Called whorl-beads, these may have been spindle-whorls, used for spinning. More distinctly spindle whorls were terracotta pear-shaped objects found at Kodumanal, flat at one end and some with bits of iron protruding from or cemented in the central hole.[35] There is also evidence of weaving in the form of actual pieces of cloth or of cloth impressions such as at Adichanallur and Kodumanal.[36]

Analysis

Thus, the major crafts we are dealing with in the Iron Age in peninsular India are metallurgy and metalworking (of copper, iron and steel), pottery production, bead-making, spinning and weaving. While there is a tendency to conflate craft and sedentary living, theoretically most of these crafts can be practised by both nomadic and sedentary craftsmen. Mining of metal ore in many cases is practised on an almost transitional basis where miners work at the source area for brief periods of time, smelt the ore and transport refined metal back to metalworking centres or habitations. For many urbanised societies such as the Harappan, the absence of metal in the vicinity had necessarily meant the transporting in of refined metal in the form of ingots, produced either through their own expeditionary efforts or perhaps by obtaining ingots from local people who lived near the mines and worked the ore. Metallurgy practised at source areas has long been considered as the most efficient or optimum means of obtaining refined metal as the craft is a heavily weight-reducing industry, requiring large amounts of ore to produce far smaller quantities of refined metal. Other inputs like fuel too are readily available at the forested source areas. This would have been the situation for both copper as well as iron, except that iron is far more common than the former metal.

Early iron smelting furnaces too were probably not complex. The most primitive type was the bowl furnace – just a depression in the ground into which air was directed through tuyeres, the whole structure probably having a domed superstructure. The broken or powdered ore and charcoal would be mixed together and fed as charge in layers. Probably the earliest furnaces would not even have had an outlet for tapping the slag. Reduction of iron took place in the solid state, as cast iron would have required considerably higher temperatures than that achieved in this type of rudimentary furnace. Cast iron would have to wait for the invention of the blast furnace. Breaking the furnace would retrieve the bloom, produced at the bottom of the furnace resting just over the accumulated slag.[37] Early furnaces are rarely archaeologically found as the superstructure is broken to remove the bloom – what remained archaeologically would be the slag and perhaps the tuyeres as well as some portions of the furnace walls. Reconstructions of the megalithic furnace at Naikund in Vidarbha, however, indicate that the furnace walls were made of coils of clay and the whole structure was a slightly more developed version with arrangements for tapping slag.

There is also nothing unusual in the peripatetic nature of smithing – in fact, some of the earliest specialists in antiquity were considered by Childe to have been smiths. In the case of copper, objects could be made by annealing (alternate heating and cooling) or by casting and the most that would be required was a furnace with bellows and basic tools like hammers and a mould. Most iron, which cannot be cast without a blast furnace, required an ordinary furnace with bellows and forging implements. In Africa, Iran and various parts of India, the itinerant smith was very much a reality and continues to be so.

Spinning is a necessary component of a pastoralist's life and is done while on the move. Apart from the material being spun which among pastoralists is usually wool, the tools required are extremely simple – nothing but a spindle and a whorl.

Pottery and bead-making come into a slightly different category. The issue of pottery production is ambiguous as it involves the production of delicate, breakable objects. Pottery production does not, however, require completely sedentary living – it can be practised by transhumants who remain for short periods of time at particular places. There are ethnographic cases of potters who are mobile, but who live for about half the year in places other than their homes. Personal observations of tribal potters in a village in the Chhota Udepur area of Gujarat revealed that burnished and glazed pottery could be prepared and fired at a completely subsistence level within a day. This was possible partly because the equipment used was rudimentary in the absence of a wheel or other specific potters' tools (a sickle, for example, was used to scrape off extra clay) and firing was done without permanent facilities, with the bonfire method. Obviously, the raw materials needed to be available, such as clay, materials for glazing and fuel and water. It would, theoretically, not be difficult to practise such methods on the move as long as one settled for a brief period of time near clay and water sources.

Similarly, the tools required for bead-making are all portable, whether one considers those needed for the shaping of beads, or grinding or polishing or perforating. The only requirement that is physically immovable, but can very often be rudimentary, is a heating facility for stone and unfinished beads. In present-day Khambhat, however, known for bead-making at least since the seventeenth century, bead makers are completely sedentary, mostly working in organised workshops. In contrast, in nearby Nagara where beads are drilled, drillers often carry their equipment to their fields so they can drill beads at the same time as look after the crops.

However, the case of pottery production and bead-making are slightly different from that of metallurgy or spinning as the first two crafts lead to the making of in artefacts that are not absolutely essential for nomads. Nomads usually prefer leather bags or baskets or metal containers for carrying provisions rather than pottery, which is breakable. This is not to give a blanket statement that pottery is not used by nomads – Gujarati pastoralists often use pottery vessels. Beads are easily available from settled peoples or the market and hence need not be produced by nomads themselves. However, spindles and whorls are required for their own day-to-day spinning needs while metal is immediately necessary to make tools and weapons. Apart from smiths being often peripatetic, it is apparent that nomads all through history went to great extents to provide for metal tools and weapons. We must also keep it in mind that actual production by mobile societies needs more than a theoretical, or more important, a technical ability to undertake production. Probably, if these crafts had been intrinsic to the life of nomadic societies, there would have been a greater effort to directly provide for their production. Thus, perhaps one finds almost no cases of pottery production or bead-making among nomadic societies.

Let us try and correlate the evidence for craft and megalithic societies.

1. If megalithic societies were agriculturalists, there is little confusion as craft production is often associated with sedentary living. There may have been separate craft workers, perhaps some possessing certain skills or perhaps it is more likely that crafts were practised by craftsmen-cum-agriculturalists. For this, we do have the evidence of habitation-cum-burial sites with more than a meter of occupation deposit, where craft production seems to have been undertaken, be it Kodumanal or Mahurjhari with bead-making or Kesarapalle perhaps for pottery production or habitation and habitation-cum-burial sites from Moorti's list of metallurgy and metalworking sites. In most cases of habitation-cum-burial sites, such as Brahmagiri or Kodumanal, the correlation between the megalithic burials and certain occupation layers in the habitation mound have been worked out. Even so, slight dissonances exist, such as the absence of quartz beads in the megaliths at Kodumanal while abundant evidence exists for bead production in the habitation mound, or that copper bangles may occur in the megaliths at Brahmagiri but now in the occupation levels, or the fact that at most sites,

etched carnelian beads are more likely to be found in the burials than in occupation layers. Some pottery shapes, too, are limited to megalithic burials (as at Brahmagiri[38] or Maski[39]). These could be explained as burial-specific artefacts, produced for and used only in the burial ritual. Or it could also well be that absence in occupation levels was a factor of excavation techniques.

2. If, however, megalithic societies were nomadic, how then could they procure crafted objects and do they, in fact, also produce them? Let us engage in a theoretical exercise to debate this issue, keeping in mind that there may be variations in nomadic routines from transhumance to long-distance forays. Pastoral nomads may obtain needed craft objects from settled peoples, often giving in return pastoral products such as wool, leather or cheese, or products of upland forested regions that they may frequent. If we assume that megalithic folk were nomads and different from settled peoples whose remains we note in the black-and-red ware horizons of habitation sites, then the differences should show up archaeologically. Perhaps, then, the dissonances as pointed out above may become significant. At the same time, we should keep it in mind these could have been burial-specific artefacts. In a variant of this situation, sedentary craftsmen, on conquest, could be forced to produce for their nomadic conquerors.[40]

3. Nomads could also perform certain production activities when practising transhumance, particularly at places where they may settle for short periods of time. It seems that craftwork among nomads tends to focus on raw materials that are readily available to them – like animal skin or wool. Most descriptions of Central Asian nomads, for example, will indicate weaving, embroidery and felt-making as crafts practised by women. It also appears that some amount of wood-working may have been done, such as wooden utensils being made among the Dzungar of Southern Siberia[41] or birchwood arrows by the nomadic Kazakhs.[42] Craftwork may often be quite intensive as seen in weaving outputs of Qashqa'i women in Iran.[43] It has been pointed out[44] that craftwork was often a necessary component of a pastoralist's workday. In this context, the attempt was often made to simplify craft processes, largely tending towards reduction in design and technique. There may be modification in craft tools too, such that it was preferable to carry lighter and fewer tools. These could range from single tools performing multiple tasks (such as moulds-cum-crucibles among the itinerant *dhokra* casters of Eastern India[45]), unlike

their counterparts in settled societies where elaboration and expansion of the tool repertoire could be indulged in. This is immediately apparent from Wulff's description of the tools used by itinerant and settled metalworkers where the latter had a repertoire of several specialised types of files and chisels. Lucassen[46] also points out that the Hungerian Kaldarash, gypsy copper and tin smiths, preferred a traditional form of bellows that was smaller than what would have been optimum for metalworking, but prevailed with them because of their being easier to transport.

In the megalithic context, keeping in mind the peripatetic nature of smithing, as pointed out earlier, we can notice that Moorti[47] lists 14 burial sites from where there is evidence of iron smelting and working in the form of slag and other smelting remains. So also does he note burial sites in the vicinity of which there is evidence of ancient gold working. At none of these sites is there a habitation in the vicinity. It is quite possible that the working of iron or gold mines at certain places may have been of some significance, so much so that the dead were also interred nearby, almost as a marker of possession. This could, arguably, be the most important indicator for craft being practised by nomads. Vibha Tripathi[48] also considered it quite plausible that megalith builders were nomads and exchanged their metal products for agricultural and other produce.

For other crafts, we have less archaeological information. But dare it be suggested that the kinds of black-and-red ware found in some megalithic burial contexts was produced in such possibly transhumant situations? Black-and-red ware, although usually of fine texture attesting to reasonably well-levigated clay, was produced on the tournette and often appears to have been ill-fired, suggesting the use of the bonfire method. The tournette, often mistakenly called the 'slow wheel', had the advantage of a pivot and socket along with rotation over completely handmade techniques. However, it is not the true wheel as it is smaller and less heavy and thus does not have the momentum or the centrifugal force of the true wheel. But because tournettes can achieve rotation for short periods, striations will show on inner pottery walls but will not have the closely spaced regularity of the true wheel.[49] It is only sedentary societies that use the true wheel, while tournettes, in contrast, would perhaps be more portable. Megalithic pottery also appears to have largely favoured conical or pointed bases. As at Nagarjunakonda, the kinds of bases mark a shift from earlier periods where flat or flattish bases were made. Rounded bases are indicative of the beating technique, which was clearly used as dabbers have been found. Conical or pointed bases are useful for cooking purposes as their shape distributes heat evenly but would require a method for suspension.[50] However, it is unclear what megalithic burial pottery with pointed bases was used for, though apparently stands were provided for their support.

4. Another scenario could be that crafts were practised by craftsmen who accompany nomads. In this scenario, craftsmen were nomadic because their clients were. Examples are available of the Inadan who accompany the Tuareg in Africa[51] and among the pastoralists of Central Asia. Phillips[52] has also assumed that the marvellous works of art used by the Scythians may have been the work of captive craftsmen. The Dzungar rulers of Southern Siberia, too, selected artisans from amongst their captives to produce much needed things like cotton and woollen goods, leather articles, paper and guns.[53]

Conclusion

To many archaeologists, the very practice of craft implies sedentary living and some go to the extent of assuming specialisation. However, as we have seen in the above instances, this is not necessarily the case. The social context of production needs to be kept in mind – megalithic societies clearly saw the interment of iron tools, weapons and pottery as intrinsic to their burial ritual. Quantitatively, there seems to be a larger number of iron objects in burials than in habitations, which may reflect beliefs and practices that were willing to put metal objects permanently out of circulation. However, the disparity might also be a feature of excavation methods. But clearly there does seem to have been a demand for ironworkers. However, as pointed out earlier, this very fact does not make for sedentary living as nomads too make available for themselves all required goods, and not necessarily only through exchanges with settled peoples. At the same time, we should also remember that several megalithic sites were late and extended clearly into the Early Historic period, as for example, Kodumanal with its Tamil Brahmi inscriptions on pottery. We could in these contexts well be dealing with specialists. Perhaps, if one were to be sensitive to the variations possible and the contexts of these early societies, one may be able to proceed towards an understanding of the technologies available to and/or practised by the megalith builders.

Notes

1. B. Subbarao, "Some Problems in the Archaeology of South India", *Transactions of the Archaeological Society of South India 1956–57*, Madras, 1957, pp. 62–78.
2. R.E.M. Wheeler, "Brahmagiri and Chandravalli 1947: Megalithic and Other Cultures in Mysore State", *Ancient India*, No. 4, 1947–48, p. 201.
3. H. Sarkar, "Kesarapalle 1962", *Ancient India*, No. 22, 1966, p. 43.
4. R. Subrahmanyam, "Nagarjunakonda (1954–60)", Archaeological Survey of India, Delhi, 1975, p. 184.
5. B.K. Thapar, "Porkalam 1948: Excavation of a Megalithic Urn-burial", *Ancient India*, No. 8, 1957a, p. 13.
6. Ibid., p. 18.
7. V. Tripathi, "The Age of Iron in India", in S. Settar and R. Korisettar (eds.), *Indian Archaeology in Retrospect: Prehistory-Archaeology of South Asia.* Vol. I, New Delhi: Manohar, 2002, p. 295.
8. A. Ghosh, *Encyclopaedia of Indian Archaeology*. Vol. I, New Delhi: Munshiram Manoharlal, 1989, pp. 111–12.
9. R. Champakalakshmi, "Archaeology and Tamil Literary Tradition", *Puratattva*, No. VIII, 1975–76, p. 119.
10. N.R. Banerjee, "The Megalithic Problem of Chingleput in the Light of Recent Exploration", *Ancient India*, No. 12, 1956, p. 23; N.R. Banerjee and Soundara Rajan, "Sanur 1950 and 1952: A Megalithic Site in District Chingleput", *Ancient India*, No. 15, 1959.
11. Banerjee, op. cit., p. 23.
12. K.R. Srinivasan and N.R. Banerjee, "Survey of South Indian Megaliths", *Ancient India*, No. 9, 1953, p. 109.
13. K.P. Rao, *Deccan Megaliths*. Delhi: Sundeep Prakashan, 1988.
14. B. Narasimhaiah, *Neolithic and Megalithic Cultures in Tamil Nadu*. Delhi: Sundeep Prakashan, 1980.
15. A. Sundara, Comments on S.B. Deo, "The Megaliths: Their Culture, Ecology, Economy and Technology", in S.B. Deo and K. Paddayya (eds.), *Recent Advances in Indian Archaeology*. Pune: Deccan College, 1985.
16. Narasimhaiah, op. cit., Sundara, "Comments", op. cit.
17. L.S. Leshnik, *South Indian 'Megalithic' Burials*. Wiesbaden: Franz Steiner Verlag, 1974.
18. Wheeler, op. cit., p. 199.
19. Sengamedu in Vriddhachalam Taluk of South Arcot district, Tamil Nadu has a cultural deposit of about 4.2 m of black-and-red ware, towards the end of which red ware sherds start to appear. At Maski, the megalithic layer was of 1.6 m thickness of which the upper 0.3 m overlapped with the Early Historic layers See B.K. Thapar, 'Maski 1954: A Chalcolithic Site of the Southern Deccan', *Ancient India*, No. 13, 1957b, p. 11. At Kesarapalle,

the megalithic occupation was 2.4 m with a Northern Black Polished Ware sherd in the upper levels. See Sarkar, 'Kesarapalle', p. 40, these kinds of deposits do not fit in with an archaeological interpretation of nomadic communities settling for brief periods, which should show more ephemeral occupations.

20. Sundara, 'Comments', op. cit.
21. Wheeler, op. cit., p. 206.
22. U.S. Moorti, *Megalithic Culture of South India: Socio-Economic Perspectives*. Varanasi: Ganga-Kaveri Publishing House, 1994.
23. Ibid., pp. 38–41; Table 3.3. We are here assuming (but which may not necessarily be the case) that the evidence pertains to the megalithic layers of sites occupied over multiple periods.
24. Ibid., p. 14.
25. K. Rajan, *Archaeology of Tamil Nadu (Kongu Country)*. Delhi: Book India Publishing, 1994, pp. 65–66.
26. Moorti, op. cit., p. 43.
27. Sarkar, op. cit., p. 74; Plate XXVA, 5.
28. Banerjee and Soundara Rajan, op. cit., p. 40; Plate XX, 4.
29. S.B. Deo, *Mahurjhari Excavation: 1970–72*. Nagpur: Nagpur University, 1973, p. 32.
30. R. Mohanty, "Significance of a Bead Manufacturing Centre at Mahurjhari, District Nagpur, Maharashtra, India", *Man and Environment*, Vol. XXIV, No. 2, 1999.
31. Y. Subbarayulu, Report on Kodumanal Excavations, unpublished manuscript, 1988.
32. Wheeler, op. cit., Plate CXXA, 40, 41.
33. Sarkar, op. cit., Plate XXIVA, 1, 6.
34. Subrahmanyam, op. cit., p. 182.
35. See K. Rajan, "New Light on the Megalithic Culture of the Kongu Region, Tamil Nadu", *Man and Environment*, Vol. XV, No. 1, 1990, p. 98.
36. Ibid.
37. R.F. Tylecote, "Furnaces, Crucibles and Slags", in T.A. Wertime and J.D. Muhly (eds.), *The Coming of the Age of Iron*. New Haven and London: Yale University Press, 1980, p. 210.
38. Wheeler, op. cit., p. 232.
39. Thapar, op. cit., pp. 51–63.
40. J. Miyawaki, B. Cuiqin and L.R. Kyzlasov, "The Dzungars and the Torguts (Kalmuks) and the People of Southern Siberia", in J. Hermann and E. Zurcher (eds.), *History of Humanity*. Vol. III, Paris and London: UNESCO and Routledge, p. 176.
41. Ibid., p. 164.

42. A. Burton, "Trade", in Hermann and Zurcher (eds.), op. cit., p. 413.
43. L. Beck, *Nomad: A Year in the Life of a Qashqa'i Tribesman in Iran*. Berkeley and Los Angeles: University of California Press, 1991.
44. J. Menon, "Mobility and Craft Production", in R.C. Heredia and S.F. Ratnagar (eds.), *Mobile and Marginalized Peoples: Perspectives from the Past*. New Delhi: Manohar, 2003.
45. L. Horne, "Itinerant Brasscasters of Eastern India", in B. Allchin (ed.), *Living Traditions: Studies in the Ethnoarchaeology of South Asia*. New Delhi: Oxford and IBH, 1994.
46. L. Lucassen, "The Chink of the Hammer was Heard from Daybreak till Dawn: Gypsy Occupations in Western Europe (nineteenth to twentieth centuries)", in L. Lucassen, W. Willems and A. Cottaar (eds.), *Gypsies and Other Itinerant Groups*. New York: Macmillan, 1998, p. 163.
47. Moorti, op. cit., pp. 39–41.
48. V. Tripathi, "Comments" on S.B. Deo, "The Megaliths: Their Culture, Ecology, Economy and Technology", in S.B. Deo and K. Paddayya (eds.), *Recent Advances in Indian Archaeology*. Pune: Deccan College, 1985.
49. P.M. Rice, *Pottery Analysis: A Sourcebook*. Chicago and London: University of Chicago Press, 1987, pp. 133–34.
50. Ibid., p. 242.
51. D. Casajus, "Crafts and Ceremonies: The Inadan in Tuareg Society", in A. Rao (ed.), *The Other Nomads*. Koln/Wien: Bohlau Verlag, 1987; M. Kirtley and A. Kirtley, "The Inadan Artists of the Sahara", *National Geographic* 156, No. 2, 1979.
52. E.D. Phillips, *The Royal Hordes: Nomad Peoples of the Steppes*. London: Thames and Hudson, 1965. See for other contexts A.M. Khazanov, *Nomads and the Outside World*. Cambridge: Cambridge University Press, 1983 and O. Lattimore, *Inner Asian Frontiers of China*. Hong Kong: Oxford University Press, 1988.
53. Miyawaki et al., op. cit., p. 164.

5 In Search of Technology

Journey from Ai Khanum to Mathura via Taxila (200 BC–AD 300)

Suchandra Ghosh

The post-Mauryan period in Indian history can be viewed as a period which saw the emergence and gradual crystallisation of contacts between centres of power in the north-west and other regions of the subcontinent. There was an impressive movement of men and merchandise resulting in exchange of ideas and knowledge. Interaction of one kind or another between the Indo-Iranian borderlands (the river Indus forms the eastern boundary and the western borders is defined by the highlands of the Iranian plateau, it is virtually the only overland access to the Indian subcontinent from the west) and the Doab became more intense as the major urban centres of the period like Taxila in the north-west and Mathura in the Doab. They were located on the great trade route that linked the Gangetic region with the north-west. There was another important archaeological site called Ai Khanum in Afghanistan which perhaps had linkages with Taxila and further east. The productive forces achieved the highest level of development in this period. This is reflected in the archaeological findings from Ai Khanum, Taxila and Mathura which demonstrate a very rich material culture. Minor objects like beads, utensils, jewellery, tools, glass objects etc., found from these sites way betrayed examples of technological understanding in their own way. Interestingly, some texts of the period in question indicate a society where people made the best use of their technological know-how.

This article aims to make a preliminary survey of the technological situation, particularly relating to things of everyday use, of some areas beginning from Ai Khanum in Afghanistan to Mathura in the Doab, having Taxila as the meeting point between the east and west in the post-Mauryan period. The text primarily used in this presentation is *Angavijja*. The *Milindapanho* is another contemporary text which will also be referred to. There are many references relating to professions, utensils and others which convey a definite idea of technological understanding and which sometimes converge with archaeological finds. Thus with a mélange of archaeology and text, an attempt will be made to situate the technological tradition, relating particularly to minor objects, of the period and the geographical locale under review through the understanding of a practitioner of history.

We wish to begin our search from a Hellenic site called Ai Khanum in northern Afghanistan. Situated at the confluence of the Oxus and its tributary the Kokcha, this site remained active from 3rd century BC to 145 BC. The material remains of Ai Khanum on the whole give us an image of economic prosperity.[1] Its location helped its masters to tap the resources of Badakshan, rich in lapiz lazuli. Since it was situated on the bank of the Oxus, agriculture flourished. It appears that the most prosperous period of the city followed the victory of Eucratides to the south of the Hindukush. This was perhaps possible as Eucratides could extract and mobilise the resources from the areas to the south of the Hindukush. Unfortunately this economically and strategically important city came to an end with the nomadic invasion of c. 145 BC. It is important to note that most of the materials unearthed from the site carry the distinctive imprint of Hellenism.[2]

Apart from the markers of Hellenism, the materials unearthed from Ai Khanum offers us a picture of developed scientific and technological knowledge. These are evident from remains of everyday use. In this article some of them would be mentioned. We may start with the vessels found from excavation which were gaily decorated with fret work and encrustation. Fret work and encrustation requires knowledge of well developed ceramic technology. Wine presses unearthed at Ai Khanum definitely indicate local production of wine but at the same time presence of the wine press is an indicator of the existence of a knowledge of agricultural chemistry in terms of fermentation of agricultural produces to generate alcoholic beverages. It actually presupposes a knowledge not only about the proper crops suitable for fermentation but also of the control of the fermentation process itself. It is important to mention that Ai Khanum could boast of exceptional glass wares apart from ceramics. Glass blowing technique was used which perhaps is an indicator of developed technology. The finding of glass prism and other objects made of glass definitely bears a testimony to the existence of optical knowledge and practice in this particular area.[3] Goldsmiths were also extremely proficient and encrustation was a common phenomenon. The jewellery found at Ai Khanum comprise pearl necklaces, pendants with encrustations manufactured from the vast repository of glass, semi-precious stones, mother of pearl, bone and ivory. Gold plating technique was also known. This is evident from an interesting gold plated silver plaque of goddess Cybele found in the temple at Ai Khanum. In the history of western alchemy this has been identified as aurifiction or the art of faking gold. A bust of Hermes has been unearthed from the ruins of the gymnasium at Ai Khanum. This, along with the gilded plaque of Cybele may perhaps indicate a developed knowledge of alchemical culture as the art of chemistry is ascribed to the mysterious Hermes Trismegistos, or Hermes, the Thrice Great.[4] In the reports of Ai Khanum we find that the excavators have designed an instrument for preparation of fine sections of agate (or may be other mineral bearing stone) used for ornamentation. It indicates perhaps the existence of a similar kind of stone sectioning technology.[5] The excavation report gives us an idea of the knowledge of minerals at Ai Khanum. These minerals comprise not only semi-precious or precious stones but also transition metal bearing ores and these later ingredients or minerals might have found their way in the production of coloured glass objects or faience bodies. Faience is made by breaking up quartz into small grains, adding a small amount of lime and fusing until the surface of the quartz has flowed and cemented the whole into a solid mass. The quality of the faience depends upon the fineness of the grains and degree of fusing. It may be noted here that a faience head has also been found from this site. The head from Ai Khanum appears to be the largest faience sculpture known from the Hellenic world.[6] The list of minerals also include the famous pigment orpiment. It is used for metallurgy, medicine and also for cosmetics. There were ateliers of metallurgy where production and work on iron, copper, gold and silver were current. Though we do not have any artisan's zone at Ai Khanum, the nature of the iron objects spread throughout the site pointed to the presence of a very efficient class of artisans. In the arsenal, the remaining scraps of iron slags suggested that there was an atelier for fabrication of Graeco-Bactrian armaments. Of extreme interest is the discovery of four pieces of armour made of iron at the arsenal. These were parts of the cataphractus,[7] used by the Greek army in Ai Khanum to face the menace of the nomadic invaders who also wore heavy armour.

As regards Taxila, it is distinguished by its location on the great trade route that linked Gangetic region with the North West. In fact it is located on the shortest possible way from Doab to Central Asia. It was also close to the Hazara hills and so able to command the roads to Kashmir and the upper Indus valley. The site is also distinguished by its agricultural productivity and the central Asian trade added to its prosperity. According to Arrian, "it was the greatest of all the cities between the Indus and the Jhelum".[8] Later Taxila was the residence of a Mauryan viceroy and perhaps an important administrative centre of the Indo-Greek kings. Alexander's stay in Taxila was quite short and so the Greek elements in the culture of Taxila was the contribution of the later Greeks.

The cultural influence of the Greek rule, as revealed in the excavated material is quite significant.[9] The greatest influence of the Greeks is seen in the technology of crafts, domestic utensils and manufacture of jewellery of various kinds.[10] Thriving commerce facilitated the infusion of technical expertise. This is evident from the techniques of granulation and filigree which are seen in the ornaments discovered from Sirkap. The technique of granulation, i.e., the decoration of a gold surface with fine granules was known in Greece and the Near East for a long time. The most perfect examples of patterned granulation at Taxila are to be seen on the amulet containing tooth relic (no. 80) and an amulet case (no. 82), both of which date around early 2nd century. BC. The art of filigree is effected by soldering fine wire to the surface of the metal, the wire being either plain, twisted, plaited into a chain or beaded. Examples of filigree work may be seen in the disc pendants and this method of decorating gold and silver ornaments surely came from the Graeco-Roman world. Like Ai Khanum the art of encrusting jewellery with gems was also known in Taxila. Encrustation was probably known in India from the early times and it was not a borrowing of Greek culture in Taxila. The semi-precious stones used in jewellery in Taxila were carnelian, chalcedony, agate, onyx, garnet, jasper, lapis lazuli, rock crystal and others. It is understandable that inlaying of gems in ornaments required technological knowledge. What is also noteworthy is that the semi-precious stones used in Taxila for encrustation were the same as those used in Ai Khanum. High technological efficiency in metallurgy has been noticed by Marshall in Taxila. Bronze containing 21–25 per cent tin, with little lead but some antimony, was in general use in Taxila for casting domestic utensils, bells, ornamental pieces. In the opinion of A.K. Biswas, "the Taxilians preferred high tin bronze presumably on account of its easy melting and casting properties".[11] The metal was also easily available from Afghanistan. Interaction with the Grecian world also helped in the development of technology. An important example was an unguent flask obtained from the Greek stratum of Sirkap. Its shape, material and technique indicate its Hellenistic character(flared base, long neck and wide projecting brim). It contains large percentages of tin and lead. The advantages of adding lead was that it made the metal stronger and easier to cast. As early as 4th century BC a technological breakthrough in Taxila was made and that was the earliest evidence of the use of zinc in the world. From Taxila we also learn about the copper-nickel alloy. It was evidently prepared by the smelting of mixed ores of copper and nickel which are known to occur naturally in the province of Yunnan in China and must have reached India through trade.[12] This alloy gave a durable silvery lustre and was used for minting coins by the Indo-Greek kings Euthydemus II, Agathocles and Pantaleon. Perhaps nickel alloys of this kind have not so far been found at any other ancient site in India. As regards the bead industry, it has been seen that there have been technological development in the bead industry during the early historical period and Taxila was no exception. Beads of glass, shell, faience, semi-precious stones, coral etc., have been unearthed. Metal wares of Taxila are also very impressive. Sirkap for the first time produced evidence for the use of copper, bronze and silver for household objects. The copper and bronze caskets were all cast and sometimes embossed. Bowls, cups, ladles, spoons, dishes and plates were all made in copper, bronze or lead.[13] Silver wares show the taste of the class of people who lived in Sirkap. Taxila has also produced evidence of glass and they were used for bangles as well as vessels, among which those of particular importance were the translucent flasks. With this brief overview of select finds from Taxila, it is at least understandable that Taxila could boast of a technological knowledge which was a combination of indigenous and Hellenic wisdom.

Our last destination is Mathura which was from an early age a natural and convenient area for human settlement. Mathura emerged as an urban centre only in the 3rd century BC. The rationale behind the choice of Mathura lies in the fact that Mathura is distinct from the other localities of the Ganga valley as it became an integral part in the political changes that were affecting north-western and western India from the second century BC. Right from the rule of the Indo-Greeks to the Kushanas, Mathura was always connected to the then centre of political gravity, i.e., the north west and was also linked with the Delhi–Aravalli Axis and the Cambay node.[14] Thus this linkage provided some commonalty in the material culture of Mathura with those of the north west. An important site very close to Mathura was Sonkh.[15] Numerous etched beads were found from excavation. These beads were made in agate, carnelian and chalcedony. Spherical beads etched with minute spots or dots all over the body were quite popular at Mathura in the first two centuries of the Christian era. This decoration is also found in the beads from Taxila. While commenting on the etched beads found from Mathura, C. Margabandhu opines, 'it appears from the Mathura finds that the lapidarists exercised judicious selection of the material within their reach and improved on the natural form of the minerals by careful shaping with a sense of proportion and polishing. The bead maker also exhibited a keen sense of decoration and great deal of skill in the manufacture of etched beads of chalcedony, agate and carnelian.'[16] Among the potteries of Mathura, mention may be made of the Kushana glazed pottery. The glaze is a blue-green copper glaze based on lead with copper and iron combinations as colouring agents. From Sonkh a large number of metal objects have been unearthed. The metals are copper, iron, lead and brass. Those of copper and iron appeared through all the periods. The finds comprise weapons, tools, other implements, household and ornamental objects.[17] An interesting find from Sonkh are the spindle whorls. Terracotta spindle whorls of different shapes have been found mostly in the Kushana period.[18] It tallies with the image of Mathura as an important textile centre. The important antiquities of Mathura comprise games men, decorated discs, bangles and beads of semi-precious stones and bone, bone arrowheads, stone pestles and a variety of copper and iron objects. Luxury items include a comb and stand of ivory, soapstone caskets, varieties of semi-precious stone beads and gamesmen in the form of tiny animals. Mathura was not very lucky as regards natural resources and agrarian produce like Ai Khanum and Taxila but this was compensated by its nodal position on trans-regional trade routes. The Kushanas exploited the nodal position of Matheura and it became an important centre of production for cotton, gold, iron and stone goods.

It may be said that craft and technology go hand in hand, however minimum may be the use of technological know-how and a look at the inscriptions from Mathura will acquaint us with the different craft specialists of the region. They were jeweller *(manikara)*, goldsmith *(suvarnakara)*, worker in metal *(lohikaraka)*, iron monger *(lohavaniya)*, dyer *(rayaka)* and so on. As these people were encountering people from other areas regularly, they perhaps made the best use of their indigenous knowledge as well as knowledge gained from outside influence for their end products. Thus in this period Mathura became very prosperous leading the author of *Lalitavistara* to comment *'iyam Mathura nagari, riddha cha sphita cha kshema; cha subhiksha chakirna bahujanamanusya cha.'* It will perhaps not be out of place to mention here that the same Mathura was described in the Anguttaranikaya in the 6th century BC as a place where the ground was uneven and dusty and alms were hardly obtainable from people.[19] Thus this complete transformation of Mathura was possible only with trade and urbanisation which again was linked with technological understanding.

Now we shall turn our attention to a Prakrit text called *Angavijja*[20] belonging probably to the Kushana period or a little later. It must be categorically stated that this is no attempt to match the literary image with the archaeological evidence briefly mentioned above. The text is used here in order to suggest how technological understanding of a society is reflected in textual imagination. *Angavijja* is a text which belongs to western India written in Maharashtri Prakrit. As in the post-Mauryan period north-west India up to Gujarat fell within an ethnic interaction zone referring to *Angavijja* would probably not be illogical. In spite of the cultural plurality in the geographical locale of our choice, there was some sort of homogeneity in the region which fostered a spirit of interaction and mingling among the people.

In the 28th chapter of *Angavijja* we find a long list of professions. Without going into the details of the number of professions mentioned, what is fascinating is the micro level distinction made between some of the professions. Thus we have lac dyemaker *(alittakakara)*, dyer *(rayaka)*, dyer specialising in red *(rattarajjaka)*, lacquer worker *(jatukara)*, all known by their distinct specialisation which definitely needed specialised knowledge. Again silk weaver *(kosejja/kosa kara)* and bark-fibre weaver *(vaga)* are separately mentioned indicating use of different technology for the weaver's craft. Apart from the mention of bronzesmith, ironsmith, utensil maker, metal workers also include lapidary *(manikar)* and inlayer *(kottaka)*. Though the references to other metal workers are quite common in early India, inlayer *(kottaka)* is not always listed among the crafts person. However that the technology of inlaying was known is evident from the finds of Taxila and Ai Khanum. In the domain of textile (chapter 31), *Angavijja* makes interesting references. Apart from mentioning basic textile materials and the knowledge of gold brocade *(suvanna patta)* and tinsel printing *(suvanna khasita)*, it refers to a kind of cloth called *'loha jalika'* which must have been a kind of warrior's dress and could be partially equated with the cataphractus that we have found in Ai Khanum and see in the Scytho-Parthian coins of this period. Production of *loha jalika* surely needed technological knowledge and reference to it in this text points to the acquaintance of the author of *Angavijja* with this kind of warrior's dress. Granulation was also known. It refers to a class of bowl called *panasaka* which had its outer body granulated like that of a jack fruit.

Another text that we would like to refer to is called *Milinda Panho.*[21] Information supplied by the *Milinda Panho* regarding technological knowledge is minimal. But the objective behind referring to this particular text is its contemporaneity with the period discussed here along with its geographical locale, i.e., north-west. It describes the city of Sakala which has been identified with Sialkot, now in Pakistan. *Milinda Panho* while talking about metal craftsmen present in that city, excluding blacksmiths, states that they could manufacture articles from gold, silver, lead, tin, copper alloys, iron and even knew how to debase metal for use. It also makes minute differentiations between professions, e.g., bow manufacturers and bow-string makers are placed in different categories. Other professions mentioned are dyers, dye manufacturers, cotton-thread spinners, carvers in ivory and many others. It was a city where people from different places converged thereby resulting in exchange of ideas and definitely expertise. The shops are described as full of silver, bronze and stoneware which were probably locally made. There is an indication that the shops were filled with an abundance of alluring jewels and it was frequented by crowds of elegant merchants. Thus the image gathered from the representation of Sakala is that the people of the city must have had an idea of the technological tradition present in that period.

Thus archaeological material, substantiated and supplemented by literary references indicate remarkable technological know-how relating to chemical practices, metallurgy, metal working, production and colouring of glass, making of weapons, agricultural implements, domestic utensils, jewellery and much more. This was perhaps possible because of free movement of people, trade and ideas. Another important factor behind technological advancement in the area under review was definitely its multiprong linkages of which the Central Asian connection played a significant role. People of the north-west were themselves of varying cultures and the region displayed networks of wide ranging territorial and inter-civilisational contacts. Availability of raw materials and location on the trade routes were an added advantage. The Gulf of Cambay also provided another important linkage. According to the Periplus of the Erythraean sea coins of Apollodotus and Menander were in circulation in Barygaza. Again we have coins of Strato, Menander, Antimachus and Apollodotus at Mathura which is an indicator of commercial links between Mathura and Barygaza.

Cambay was the centre of lapidaries' art and it had a ready market in Barygaza. The political authorities ruling in this region in the post-Mauryan period were the Indo-Greeks, the Scytho-Parthians and the Kushanas. Kshatrapa rule was witnessed at Gujarat and Mathura for a certain period of time. Multiple political powers did not act as a deterrent to the exchange of ideas and movements of men. Uninterrupted interaction between different areas could be noticed irrespective of political authorities. The traders who are said to be the purveyors of culture had some sort of autonomy of their own which is clearly reflected in the profusion of guild like organisations in this period. So one does not perceive any break in the continuity of the flow of ideas and exchange of technological expertise with changing political authority. Nevertheless when a vast region remains under the control of a single political authority it is natural that the situation becomes more favourable. This was applicable to the Kushana rule which extended from Central Asia to the Ganga valley for a substantial period of time and thus offered political unity. This political unity was congenial to resource extraction, mobilisation and utilisation of these resources for technological development.

Notes

1. P. Bernard and C.I. Rapin, '*Campagne de fouilles a Ai Khanoum: III. Le Palais. La tresorerie*', *BEFEO* 68, 1980, pp. 6–38.
2. Suchandra Ghosh, 'Hellenism at Ai Khanum', *Journal of the Asiatic Society*, Vol. XLVII, July, Kolkata, 2005, pp. 77–86.
3. Somnath Ghosh and Suchandra Ghosh, 'Not by Archaeology Alone', *Archaeology and History: An Interface for Holistic Study of the Past*, School of Studies in Environmental Radiation and Archaeological Sciences, Jadavpur University (forthcoming).
4. Somnath Ghosh, 'Alchemy or Proto-Chemistry of East and South-East Asia and that of the Greco-Arabic and European World: A Comparative Study', in Proceedings of the First Biennial Conference of Indian Association for Asian and Pacific Studies, Kolkata, 2004, pp. 484–500.
5. Claude Rapin, *Fouilles D'Ai Khanoum*. Vol. VIII, Paris: Boccard, 1992, p. 68.
6. O. Bopearachhi, 'A Faience Head of a Graeco-Bactrian King from Aikhanum', *Bulletin of the Asia Institute, Alexander's Legacy in the East, Studies in Honor of Paul Bernard* 12, 1998, pp. 23–30.
7. Fr. Grenet, J.C. Ligeret and R.D. deValence, 'L'Arsenal', *BEFEO*, Paris, 1980, pp. 61–63.
8. G. Fussman, 'Taxila: The Central Asian Connection', in Howard Spodek and Doris Meth Srinivasan (eds.), *Urban Form and Meaning in South Asia:*

The Shaping of Cities from Prehistoric to Precolonial Times. Washington, DC: National Gallery of Art, 1993, pp. 83–100.

9. Sir John Marshall, *Taxila*. Vol. II, Cambridge: Cambridge University Press, 1951, p. 618.
10. A.H. Dani, *The Historic City of Taxila*. Paris: UNESCO, 1986, p. 86.
11. A.K. Biswas, *Minerals and Metals in Ancient India*. Vol. I, New Delhi: D.K. Printworld, 1996, pp. 255–80.
12. Sir John Marshall, op. cit., p. 571.
13. Ibid., p. 125.
14. B.D. Chattopadhyaya, 'Mathura from the Sunga to the Kushana Period: An Historical Outline', in Doris Meth Srinivasan (ed.), *Mathura: The Cultural Heritage*. New Delhi: American Institute of Indian Studies, 1989, p. 19.
15. Herbert Hartel, *Excavations at Sonkh: 2500 Years of a Town in Mathura District*. Berlin: Dietrich Reimer Verlag, 1993.
16. C. Margabandhu, 'Etched Beads from Mathura Excavations: A Note', in Doris Meth Srinivasan (ed.), *Mathura: The Cultural Heritage*. New Delhi: Manohar, 1989, p. 202.
17. Herbert Hartel, op. cit., p. 215.
18. Ibid.
19. B.N. Mukherjee. *Mathura and its Society*. Calcutta: Firma K.L. Mukhopadhyay, 1981, p. 132.
20. Muni Punyavijaya (ed.), *Angavijja*. Vol. I, Varanasi: Numismatic Society of India, 1957.
21. V. Treckner (ed.), *Milindapanho*. London, 1928.

6 Architecture and Technological Advancement

A Case Study of Select Forts of Deccan

M.N. Rajesh and M. Ramakrishna

The present article begins with a general sketch of historiography in India which relegated scientific achievements to the background. It then focuses on the forts of Deccan and tries to show that there were constant innovations in the field of technology in medieval Deccan based on archaeological reports, particularly in the area of defence architecture.[1]

The Problem

Europeans discovered China during the Enlightenment and India during the Romantic period. They are therefore predisposed to find science in China and religion in India.[2] The above statement captures many facets of Indian history that were relegated to the background in favour of a past that was credited with an abundance of spiritual achievements and which was a staple of the nationalist historians and the orientalists. In post-colonial India, this trend has been challenged but there still remain many dark areas. Medieval archaeology is one such area since the primary focus of archaeological research in India has concentrated a lot on the pre-historic and the proto-historic periods, large gaps remain in the understanding of medieval India.

In this context it would be worthwhile to begin by pointing out that the modern day states of Karnataka and Maharashtra have the largest number of forts in India and together with the Gangetic plains and Rajasthan, account for more than thousands of forts. Yet there is surprisingly a great paucity of both primary and secondary literature on forts which is a grey area leading us to many generalisations that are shaky, often pliable to myth making like the construction of the image of Shivaji in different oral and written traditions. From the 19th century, in the Maharashtra public sphere, Shivaji has been constructed and reinterpreted again and again and we see two types of sources, one called the *bakhars* or the official records and the other, the *powdas* sung by the *shahirs* as part of the folk tradition. A study of forts would bring to us the material basis of such myth making and how such exercises seem historically inaccurate.[3] It is therefore clear that the material co-relation with textual evidence would place us on a sound footing and at the same time help us to broaden our understanding of historical phenomena.

Keeping the centrality of technological change and the problematic of history writing certain aspects are taken up for analysis with the forts of Deccan as an anchoring point. One is chronology and the characterisation of periods, the second is the problem of social control which is based on the analogy that as theories of patriarchy may help us to understand household oppression, to understand large-scale control of populations, the role of structures like forts need to be studied. It is here that the notion of the study of military history is felt which would explain to us in clear terms, whether we are right in using concepts like force since large-scale populations were maintained by coercion and not by consensus, if we are to go by that formulation.[4]

There has been a periodisation of Indian history since the British times into Hindu, Muslim and British periods, and a construct that led to the production of a whole gamut of literature based on the communal identity. After independence, this categorisation was dispensed with and a new categorisation namely ancient, medieval and modern period that was free of any communal labels became the standard practice. However, AD 1200 was the date that marked the onset of the medieval period and equated the medieval period with the rise of the Muslim dynasties in India, implicitly. Thus, the futility of renaming was only an exercise in labelling without any qualitative indicators. Since then the debate has dragged on and the redefinition of the 'medieval' has occupied considerable space. A remarkable contribution in this direction is the work titled *Interpreting Early Medieval India* by B.D. Chattopadhyaya which argues that the transition from the early historic period to the early medieval period was marked by structural changes in the field of production characterised by large-scale agrarian expansion.[5]

Similar parallels are reflected in the history of Islamic architecture and the division is five-fold – the early period, the classical period, the medieval period, the age of the gunpowder and the modern period. Though this periodisation may not suffice to explain fully the medieval period in the Indian context, the fact that an overarching periodisation for the discipline of history may not make full sense as it subsumes many areas or sub-disciplines like art history, military history, history of the family etc., all under one encompassing rubric of a dominant theme. Political history was hitherto the dominant theme and presently socio-cultural history is on the centrestage.[6] In such a methodological dilemma, the investigation of forts would shed light on the need for periodisation. Therefore this topic investigates the forts of Deccan and among other themes, tests the validity of periodisation.

There are voluminous references to the relentless campaigns of Aurangzeb who spent a good part of his life in the Deccan and the recalcitrance of the Deccani generals who gave a tough opposition to the militarily advanced Mughal forces.[7] The question of chronology and relevance of technology can be paraphrased in a single quote "Much of the problematic issues attending the reconstruction of the history of medieval India disappears or thins down provided the working historians succeed in freeing their mind from the myth of 'Dark Age' associated with the effective presence of Islam in northern India".[8] A similar view is echoed by André Wink when he talks of the large-scale proliferation of urban centres under the Mughals and that the cities always included large number of non-Muslims.[9] This factor is important for the simple reason that technology in medieval India has been viewed trying to "relate it with the 'rise and fall' of Hinduism, Buddhism and Islam and are bound to reap poor historical harvest …".[10] It arises from a wrong reading of the sources that is again anchored on the view that the indigenous forces that had been at work in India disappeared with the advent of Islam and that the ways of the rulers could supplant those of the numerically superior spread over the length and breadth of the country.[11] A classic example is of Benares city that continued to exhibit, as quoted by André Wink "Here we do have a city with ancient origins, a vestige of Hindu urbanism that by the seventeenth century could still be described as "a large and very well-built town, the majority of the houses being of brick and cut stone, and more lofty than those of other towns of India."[12] Sources therefore tell us only part of the story leading us to the assumption that the history we know is only a series of accepted judgements and with a small digression it may be pointed out that the recent explosion of interest in the archaeology of warfare has demolished the notion of the peaceful savage and in turn has spurred equally important findings based on novel readings of sources as disparate as the study of tombs in pre-modern India.[13]

The ecological context determines the settlement pattern and the production process along with the political structure to a very large extent. As a physiographic region, the Deccan has a rocky profile with abundant availability of rocks, proliferation of many hillocks all of which provide a suitable setting for the rise of forts. The state of Maharashtra is one of the states in India with a very large number of forts on account of the terrain, i.e., the hillocks propped up by the Western Ghats. The adjoining states of Karnataka and Andhra Pradesh also have very large number of forts, well-known among them are the forts of Golconda, Bidar[14], Gulburga and Bijapur to name a few. The above mentioned forts were also the strongholds of the Qutub Shahi (ard later the Nizam Shahis), the Barid Shahi, the Bahmani and the Adil Shahi kingdoms. Their fortunes depended to a large extent on the defence mechanisms in these forts and the political vicissitudes were clearly reflected in the changing overlordship of the fortress. We therefore witness many innovations in the forts of Deccan that show the adoption of newer techniques and the primary concern was defence and water supply and is best illustrated in an advanced stage in the fort at Bijapur. Here the impact of the large-scale use of gunpowder due to the spread of local techniques of gunpowder preparation and also the fact that the Bijapur fort is located in an open field, called for more developments. The walls are high and thick and of ashlar masonry. The lime mortar used is of very tenacious nature and the bastions have provisions for mighty guns.

Mud Fort with Additional Mortar Constructions

Empire building and the 'building' or 'rebuilding' of forts (for want of a better term) went hand in hand. This theme needs to be investigated again for two reasons in the context of Deccan: the fall of Vijayanagara and the post-Vijayanagara period has been portrayed as a period of the decline of the Hindus and by extension the failure of the militarisation of the Hindus. This argument assumes that Vijayanagara was the representative of the 'Hindu' in medieval Deccan. On the contrary, we see elements of borrowings amongst all the powers of Deccan in terms of all cultural aspects. To give a small example, the Vijayanagara sculptures make a significant departure from the traditional temple architecture and portray the bearded Portuguese horsemen and Muslim archers who are also depicted in contemporary literature.[15] Likewise, the use of the Islamic arch that was not found in the earlier dynasties is seen in the Vijayanagara period. In the forts of Golconda, such arches are commonplace and clearly show that they are additions of a later period, recognisable on account of two features:

1. The arches are not in conformity with the original plan and their arrangement is a later addition reflecting a change in the architectural programme of the structure that was being constantly modified.
2. They signify the shift from a purely trabeate system of building that used only stone blocks to a combination of arcuate and trabeate method wherein there was room for more shapes. This development was possible only with the introduction of mortar

Improvisation in Mud Forts at Narkhuda to Enable Use of Muskets

technology introduced in India during the Sultanate period. (Mortar technology also saw the emergence of the Turkish bath or the hammam in forts which are clearly datable from the 13th century AD, thus helping us to evolve typologies based on technological change and style.)

To look for the rout of Vijayanagara purely in the domain of military strategy would underplay the role of the political, a point emphasised by Richard Eaton, who argued that it was the overambitious attitude of the last ruler to have a vast empire which would enhance his prestige which led him to incorporate all territories from Kalyana in Bidar, the northernmost part in Tirumala that provoked a backlash and united all the Deccani powers against the Vijayanagara. This settles once for all that the fall of Vijayanagara must clearly be seen in the realm of the 'political' and not as a military defeat. One point which emerges from this is that this opinion has become pervasive in the communal portrayal of history and misses some well-known facts like the adoption of many Islamic motifs in the building technology of the Vijayanagara like the Lotus Mahal[16], an octagonal building built using mortar and also the presence of many shrines devoted to Muslim holy men who were worshipped by the archers employed in the service of the Vijayanagara army. Therefore the notion of Vijayanagara as a Hindu kingdom starts from a flawed premise that is not supported by contemporary findings in architecture. Other points adding weight to the above argument are as follows.

The post-Vijayanagara period acquires significance and in north India, signifies the consolidation of the mighty Mughal Empire whose hegemonic forays into the Deccan drag them into a long drawn struggle. The decline of the Mughals in the 18th century was seen as a dark age in the history of India when the Centralised Empire that covered a large part of the subcontinent was a pale shadow of its former self and ultimately disappeared. This view no longer holds good and the notion of the 18th century as an anarchic age is dispelled citing the examples of all round creativity in the regional centres like Avadh, Hyderabad, and Bengal etc., signifying the shift from a centre to regional units. Among all the states that survived the longest was Hyderabad under the Asaf Jahis who ruled up to 1948 when they were incorporated into the Indian Union.

Correlation of a few forts of Maharashtra that were in continuous occupation from an early period signifying a long historical past to the period of the Marathas brings out the point that the physical features of the forts based on the archaeological reports do not reveal much trace of their long historical past whereas literary evidence is to the contrary. The list of the forts follows.

Nandurbar, Daulatabad[17], Bhamer, Dusana, Khanapur, Khoparli, Shahada, Talodahanumangad, Hatgad, Kachana, Karhe, Pimplaner, Chauler, Mulher, Salher, Raipur, Chandvad, Achala, Jawlya, Markandya, Kanhergad, Rimbak, Anjaneri, Harshagad, Alang, Bhaskargad, Patta, Naladurg, Sidhgad, Shivaneri, Junnar, Narayangad, Lohagad, Belapur, Bhorap, Vishalgad, Raigad, Singhad, Torna, Surged, Ratnagiri, Puranagad, Sathavli, Panhala, Bagni, Bahadurwadi, Machal, Achalpur, Narkhed, Sangli, Purandar, Ahmadnagar, Deogiri[18], Bidar, Gulburga, Bellary.[19]

Watch Towers Locally Known as Burzu in Telengana

Interestingly, the normative texts on forts that have been quoted by authors like V. Ramachandra Dikshitar and also by later writers on forts like Virginia Fass do not match with the contextual reality. Among other examples is the *Ajanapatra*, a document from Maharashtra on the functioning and administration of forts. But contemporary literature is at a considerable distance from the ideals mentioned in the *Ajanapatra* demonstrating to us the futility of operating with ideal categories since the medieval period in Indian history was very dynamic.[20] In contrast to the above there is a different finding which shows that the practice of using some recorded material on the literature of forts linking the same in addition to the findings of the descriptions available, correlating it with the techniques of warfare, has been a standard practice to describe the military set-up in the medieval period. This has serious limitations and is fallacious. It is also tied up with the thesis of stagnation in medieval India which has been eroded considerably and in the case of forts we have special reason taking technology as a marker, for in the medieval period we see two groups of people, the trader and the warrior who were mobile and were also the first to adopt change on account of pragmatic reasons – the trader did not want to be poor and the warrior did not want to die and such a situation would befall them if they operated with normative categories.[21]

One fact that is central and greatly obscured by this methodology, is the total absence of any reference to the mud forts in literature, commonly referred to as *hude* in Kannada.[22] It has been conclusively demonstrated by S.K. Aruni in his doctoral thesis submitted to the Deccan College of Archaeology, to the existence of many mud forts in the Karnataka region and their role in the local defence network. Such mud forts were managed by the local warrior clans and used as a refuge by the general populace and also cattle in times of raids.[23] Mud forts were present not only in Karnataka or Deccan but also in areas that had great influence of the Mughal state like Rajasthan as the excavations at the mud fort of Dholpur show. The above finding negates the premise that there is a one to one correlation in all fields of human activity.

Among the most fascinating of all the forts is the great fort at Golconda in Hyderabad for the simple reason that this is still in a relatively good state of preservation and many of the original functions can be replicated here even today though some are in disuse.[24] Two of the most important facets that have a bearing on science and technology are the use of acoustics and advanced water supply system both of which can be seen though the latter is not in an operational state.[25] Other striking aspects found here are the free uses of beams, sometimes in combinations with arches. The arches here are the typical Islamic ones.[26] Motifs of animals are also seen here reflecting the changing times and advances in architecture. Fountains were clearly popular after the Jahingiri period and their presence is seen here in a state of good preservation. Since the fountains were a marvel of technology that also aided the visual and ceremonial status, the presence of such an engineering marvel pointed to the use of art and technology for purposes that were not only defensive but matters of political prestige. Only with the fountain would the garden be complete. Since a complete garden was the replication of a paradise, therefore the fountain was an engineering marvel and a necessity adopted in full splendour at Golconda. Another probable reason is that since the Asaf Jahi's modelled themselves on the Mughals they had to adopt Mughal symbolism or go a step beyond and thus the use of the fountain[27] was adopted. The large-scale immigration of artisans to the Deccan and subsequent patronage was another factor that helped in the technological dissemination.[28]

Thus we can say after examining the literature on forts and physically visiting them is that the forts were living institutions and as such shed new light on periodisation of medieval Indian history. While stone forts are recorded and are the object of historical enquiry, the picture of warfare we construct based on them is narrow if we ignore mud forts that reflect among other things, the local martial systems and the economics of defence capabilities. Forts with masonry and brick work also continued to be built as late as the 18th century[29] which leads us to the point that the adoption of technology is not unilinear but conditioned by economy, local usage patterns etc.

Characterisation of a school of architecture as Hindu or Islamic should be restricted only to religious buildings while forts reflect an entirely different dynamics borrowing from all traditions based on pragmatics and ideology. A perusal of the list of the above forts reveal that most of them were in use in the early medieval period but the structure does not reflect any such origins driving home the point that a fort is a living institution and is therefore in a state of constant change thus obliterating earlier features.

Whichever technological features served the pragmatic and other values were retained and constantly innovated in a fort and one of the best examples of the above is the water supply and drainage system[30], by the use of clay pipes[31], underground channels[32], aqueducts[33], use of lime masonry[34], underground water channels, drains[35] and flooring with lime and plaster.[36] The diagram below illustrates the arrangement of water supply in Daulatabad fort.[37] We also see that the large-scale use of stone in the Deccan and its conversion into various forms as detailed in the photographs below show the familiarity of the artisans with the medium.[38] It was not only stone but also two other technological aspects, firstly conversion of stone into different forms which, needed a good knowledge of measurement[39] and secondly the knowledge of optics as the structures we have detailed below have calligraphic elements which presuppose an acute sense of optical observation, otherwise there would be parallax error[40] or asymmetry which is not found in the jails or niches of Bidar fort. On the contrary the fort displays a setting of niches that is pleasing to the eye and is also continuous.

Thus we may say that technological change was very selective and depended on factors like patronage, local economy and it is futile to look at normative texts like the *Ajanapatra* as there are diversities and silences especially in the case of the mud forts. What then explains the similarities in the forts of Deccan and Maharashtra and interestingly also in the Vijayanagara and Bahmani period structures is the concept of culture area.

Notes

1. We are thankful to the University of Hyderabad for providing the necessary grant for this study under the scheme of University with potential for excellence.
2. Frits Staal, 'Introduction' to the section on Indian Sciences in Gavin Flood (ed.), *The Blackwell Companion to Hinduism*. Oxford, UK: Blackwell Publishers, 2003, p. 346.
3. Daniel Jasper, 'Commemorating the "Golden Age" of Shivaji in Maharashtra and the Development of Maharastrian Public Politics', *Journal of Political and Military Sociology* 31, No. 2, Winter, 2003, pp. 215–30.
4. Jeremy Black, 'War and the World 1450–2000', *Journal of Military History* 63, No. 3 (July 1999), pp. 669–81.
5. Brajadulal Chattopadhyaya, *The Making of Early Medieval India*. Delhi: OUP, 1996, pp. 1–26.
6. Even in political history, the theme of Islamic elements in Vijayanagara is only now being studied since the earlier works like B.L. Saletore has established a strong opinion of projecting Vijayanagara as a 'Hindu Empire' which went unchallenged. Such a view also foreclosed any option for looking at architectural developments from a frame of reference free of religious labelling.
7. J.F. Richards, *Mughal Administration in Golconda*. London: OUP, 1975 is a standard reference work on this subject which argues that the Mughal generals tried to bring about standard Mughal administrative practices in the Deccan which led to a well developed administration and paved the way for the emergence of the Asaf Jahis. It is here that the adoption of weaponry and building technology is one important aspect that explains the adoption of technology and social control and points to the long standing existence of the Asaf Jahi dynasty.
8. D.P. Chattopadhyaya, "Foreword" in A. Rahman (ed.), *History of Science, Philosophy and Culture in Indian Civilization*. Delhi: OUP, 1998.

9. André Wink, 'From the Mediterranean to the Indian Ocean: Medieval History in Geographic Perspective', *Comparative Studies in Society and History* 44, No. 3 (July 2002), p. 433.
10. D.P. Chattopadhyaya, op. cit., 5, 'Foreword'.
11. Ibid.
12. V. Ball and W. Crooke (eds. and trans.), *Travels in India by Jean-Baptiste Tavernier* (orig. French ed. 1676), 2 Vols., New Delhi: Munshiram Manoharlal, 1977, I, pp. 65–67 in André Wink, op. cit., p. 427.
13. Roberta Gilchrist, 'Introduction: Towards a Social Archaeology of Warfare', in *World Archaeology*, 2003, 35: 1–6.
14. Bidar was a very dynamic fort in terms of architectural development as the following photographs show and thus reiterates the point that since there was constant development, it was not possible to find evidence of the same in literature. The same is the case with Belgaumand Mundgod see *Indian Archaeology: A Review*. 1984–85, pp. 25–29 and p. 35.
15. *Indian Archaeology: A Review*. 1975–76, pp. 20–21. See also Anila Vergeshe, *Vijayanagara*. New Delhi: OUP, 2004, examines in detail from a fresh perspective the developments in Vijayanagara in a chapter on architecture.
16. *Indian Archaeology: A Review*. 1990–91, pp. 27–28.
17. Stewart Gordon, 'Forts and Social Control in Maharastra', *Journal of Modern Asian Studies* 13(1), 1979, pp. 1–17.
18. M.S. Narvane, *Forts of Maharastra*. New Delhi: APH Pub. Corp., 1995, pp. 42–49.
19. *Indian Archaeology: A Review*. 1992–93, pp. 38–41.
20. Stewrat Gordon, op. cit., see also V.R. Dikshitar, *War in Ancient India*. Delhi: Motilal Banarsidass, 1987 reprint, pp. 14–38 and Virginia Fass, *Forts of India*. London: Harper Collins, 1986.
21. David Nicolle, 'Medieval Warfare: The Unfriendly Interface', *The Journal of Military History* 63, No. 3 (July 1999), pp. 579–99.
22. S.K. Aruni, Mud Forts of Karnataka, unpublished PhD thesis, Deccan College, Pune, 1998.
23. *Indian Archaeology: A Review*. 1992–93, p. 134.
24. Ibid., 1973–74, p. 5.
25. Ibid., 1974–75, pp. 2–3.
26. Ibid., 1972–73, pp. 1–2.
27. Ibid., 1973–74, p. 5.
28. A. Rahman, 'A Perspective on Technology in India', in D.P. Chattopadhyaya, op. cit., 5, p. 243.
29. *Indian Archaeology: A Review*. 1992–93, p. 5.
30. Ibid., 1976–77, pp. 48–49. Ibid., 1985–86, p. 41.
31. Ibid., 1976–77, pp. 48–49. Ibid., 1973–74, p. 5.

32. Ibid., 1986–87, pp. 38–39.
33. Ibid., 1988–89, p. 40.
34. Ibid., 1991–92, p. 49.
35. Ibid., 1955–56, p. 41. Ibid., 1981–82, pp. 25–26.
36. Ibid., 1979–80, p. 33.
37. F.J. Varley, 'On the Water Supply of Hill Forts in Western India', *The Geographical Journal* 40, No. 2, August 1912, pp. 178–83.
38. A. Rahman, op. cit., 5, p. 253.
39. S.A. Paramhans, 'Units of Measurement in Medieval India and their Modern Equivalents', *Indian Journal of History of Science* 19(1), 1974, pp. 27–30 and 29–36.
40. A. Rahman, op. cit., 36, p. 253.

7 A Social History of a Minor Technology

The Tradition of Bronzes, Brass, Bell Metal and Dokra Manufacture in Bengal

Nupur Dasgupta

We are looking into the subject of minor metal works practised in rural, suburban and tribal environs in the state of West Bengal in a historical context. The products may be categorised into two broad components: (a) effigies/figurines/cast images, and (b) utensils. A very minor amount of jewellry is also to be found. Of these, the first category is marked by two distinct orientations: rural/urban and tribal, while the second is generally produced by the suburban–rural groups. The temporal context begins with the medieval period and continues to the present. Although we are not in a position to assert a theory of continuity or evolution of the monolithic technology, yet, a composite perspective of a continued tradition may be attempted.

Metal craft had formed a part of thc rural and tribal scene in the Bengal since early times. Historical data bears this out. History has recorded the classical tradition of image crafting meticulously so that we know about the Pala–Sena bronzes, while there is no record for the beginning of the tribal castings, apart from some copper hoard items from neighbouring regions of Bihar, Jharkhand and Chattisgarh, which had been parts of the historical greater Bengal. The artisan community, however, feature as a occupation – caste quite prominently in the early medieval literature.

A brief survey of the entire gamut of the minor metal craft from the medieval times onwards, reveals a complex picture with multiple stylistic, technical and social paradigms. A thorough study would need a lot of time and multidisciplinary approach. This article makes an attempt to introduce some of the major aspects of research work being carried out at present with the sponsorship of the Indian National Science Academy, New Delhi.

History and Tradition of the Technology

The minor metal handicraft tradition of Bengal is rich and typical to the environment of the region. It expresses the regional genre perhaps the best through a variegated fabric of the metal craft like the brass or *bharan* utensils, images of religious deities, and the *dokra* products, two very distinct craft styles. Many of these handicrafts have their origin among the tribal population in the regions of Bengal including Bangladesh, Bihar and Orissa, the three states that had a natural ecological bond and were a single province in the medieval and the colonial periods. The craft traditions could enlighten us as to the historical process the region underwent with regard to the progression of technology at the popular level.

The tradition of minor metal craft is most spectacularly observed in the form of the ancient Pala period bronze images. This is the most authentic point of initiation into minor metal image crafting in the history of the region.

The Pala–Sena bronze images have been objects of serious and interesting studies. The tradition of *astadhatu* images extended up to Jhewari in Chattagram district of the present Bangladesh, in the east. The main centres where the metal image crafting flourished were Nalanda and the Kurkihar Buddhist sites in Bihar, while, Banpur in Puri district, Orissa formed the southern extension where the tradition could be located. At present many of the early medieval and medieval examples of bronze images of Buddhist and Brahmanical deities have been listed as exhibits and stored items in the Indian Museum, Kolkata and the Dhaka Museum and the State Museum of the Directorate of Archaeology, West Bengal.

Most of the castings were done in the *cire perdue* or lost wax method, which has been mentioned in the early medieval Sanskrit texts.

Manasara, a 6th–11th century AD text on architecture, records[1] the tradition of casting metal images according to set rules:

"*Agastya, Kasyapa, Brigu, Gautama, Bhargava* and *Galava* are the six sages who have explained from the ploughing of the temple site (*Karshanam*) to the worship of images (*Archana*) in the finished temple, the science describing this process in the books, namely, *Visvakarma, Saiva, Pasubotyan, Kalamukha, Mahavrata, Vamana* and *Bhairava*. If with help of and in accordance with the methods here set forth in respect to the construction of temples, the making of images is faithfully adhered to, it will lead to wealth and property but deviated from there will cause poverty and distress. However, if certain useful formulae and procedures not described here are found in similar texts dealing with this process, there is no objection to the *sthapati* adopting them for his own improvement." After this admonition comes the actual description.

> The *sthapati* and the *Sthanakas* are first to prepare the wax. The actual manner of this preparation is not described). "All kinds of images, temporary or permanent, stationary a movable, are moulded in wax. Some part of the image is covered with a thin copper leaf (*tamra-patra*) and the wax is laid on two or three angulas (a unit of length measurement) deep. All kinds of images, temporary or permanent, stationary or movable, are moulded in wax. *Mulika* is spread about the part with wax. The idol is heated after it has been besmeared. If the master likes, the smearing may be done with melted iron/metal too. One half of the image, which is not covered with earth is washed in water. The process is repeated several times. If any of the minor limbs be lost through this process, the image should be furnished with it again after having been heated. But if the head or the middle of the body (*Madhyakaya*) be damaged, the whole image should be changed. If the master does not approve of the image, it should be recast." The whole process in its different stages has to be attended by many ritualistic ceremonies. The composition of this text underwent changes and is said to range between 6th century AD and 9–10th century AD. It is fairly acceptable that the *Manasara* was an almost universally known text and became the model for future treatises on the subject of art and architecture. The guidelines would correspond with those mandatory for image manufacturers from noted centres.

There are other ancient texts dealing with the technology in different manners and details: we have noted a few *Agama* texts, the *Vishnu Samhita*, the *Sukranitisara* as well as *Uttarabhaga* of *Silparatna*, and the *Manasollasa* of *Abhiliasitarthacintamani* of *Somesvara*. Prasanna Kumar Acharya,[2] Ruth Reeves[3], M.K. Krishnan[4] have all referred to these texts and our observation of the information from these sources reveal that metal craft had developed to a great extent and the theoretical documentation of this technology was also not lagging behind. In fact perhaps this technology has found a more honest and classified representation in writing than other technologies like architecture or textile production, which had also flourished greatly during the ancient and medieval times.

As P.K. Acharya points out next in his exposition on the Manasara texts: the process of casting an image appears to be much more clearly described in the *Agama* texts like the following:

> *Karanagama*[5]: "If images have to be cast in metal, the wax must first be melted and poured (out of the mould) and all defects removed with cloth."
>
> *Suprabhedagama*[6] refers briefly to the process thus: "If the images be required to be made of earth, rods (of metal or wood) must be (inserted in them), if of metal, they must first be prepared well in wax."
>
> The *Visnu-Samihita*,[7] gives a clear but brief description: "If an image is to be made of metal, it must first be made of wax, and then coated with earth; gold and other metals are purified and cast into (the mould) and a complete image is thus obtained by capable workmen."

The text of *Sukranitisara*[8] also makes a brief mention: "Images made of sand, dough, or painted, or of stucco or terracotta, or wood, or stone or metal are of relative durability in the same order." And[9] "In the case of painted images, or those made of stucco, sand, terracotta, or dough, an omission of lineaments (lakshana) will do no harm; one should beware of defects of preparation (mana) only the case of stone or metal."

But the best rendition is given in a medieval text from the Western Chalukya King Somesvara – Abhilaritarthacintamani, Manesoltasa portion.

1st Stage

> *Navatala pramanena laksanena samavrtam*
> *Pratimam Karayetpurvamuditena bichaksanah*
> *Sarvayava Sampurnam Kimcitpitam drsyeh priyam*
> *Yathoktoi ayuvairyuktam bahuvisca yathoditaih*

The initial images should be made in beeswax according to the navatala measurement (this is to be the base model). The colour would be a little yellowish (natural beeswax), the image should be beautiful and its iconography correct according to text and sastra made cleverly and with proper judgement and knowledge with all limbs according to rules and measurements.

2nd Stage

> *Tatprsthe skandadese ca krkatyam mukutye thaba*
> *Kaspuspahibham dirgha nalakam madnodbham*

Behind this image, at the shoulders or crown level, a channel should be made for pouring in molten metal. The channel should have a spouted mouth like a reed flower (Kasa) and should be like a long tunnel.

3rd Stage

> *Sthapayitva tatoscarca limpetsanskrtaya mrda*
> *majim tujamayim ghrstba karpasam satasah ksatam*
> *lavanam curnitam slaksam svalpam samyojayenmrda*
> *baratrayam tadabarya tena limpetsam pratatah*
> *etatsyyat prathamo lepam chayayam krtasosanam*
> *dinadvaye vyatite tu dvitiyah syyat tatah punah*

After putting this channel a mixture of burnt rice bush, small pieces of cloth or cotton wool and salt with clay should be prepared. This mixed clay should be put on the wax images thrice (repeatedly after drying). After the first layer is dried up in shade next layer should be applied.

4th Stage

> *tasmin suske trtiyam (yeh) tu nibido lepam isyate*
> *nutukasya mukham byaktva sarvamalepayenmada*
> *sosayet tam prayatnena yuktibhih buddhiman narah*

The last layer of clay – after the second layer dries up, the last layer of clay should be applied thickly. The spout of the channel however should be kept free. After this the whole thing should be dried carefully.

5th Stage

Now, for the amount of metals to be used for making the alloy:

> *sikyakam toulayedadou arcalagnam bicaksanah*
> *ritya tamrena roupyen hemna ba karayet tu tam*
> *siktadastagunam tamram ritidravyam ca kalpayet*
> *rajatam dvadasagunam hema syat sodasottaram*

Brass, copper, silver and gold should be weighed in the following way (according to the following measurements) – brass and copper should be of eight times the weight of wax; silver should be twelve times of: wax measure and gold should be sixteen times. The wax is probably measured along with the weight of the channel or nali.

6th Stage

Now, the process of melting the metal and pouring it:

> *mrta sambestayet yadistam kanakadikam*
> *narikelakrti musam/mujam purbbat parisosayet*
> *banhou pratapityabharcam siktam nihsarayetpurah musam/*
> *(mujam) pratapiyetpascyatpabakochchbaista banhina*

ritistamram ca rasatam navangaroirvajretdhruvam
Suvarnam rasatam yati pancakrtva pradiptitoih

The metal should be kept covered in a coconut-shaped crucible (for heating) and dried up. Then it should be heated. The wax will melt and ooze out. Then the mould should be again put on fire. For brass and copper a red hot fire is required. For silver a high fire is required and for gold even higher degree of heat five times to what is required for the other metals.

7th Stage

Casting of molten metal:

Musa Murdhani randhram lauhosalakaya
Samdamsena dhrtam musam tapayitva prayatnatah
Rasam tu nalakasyasye ksipedacchinna dharaya
Nalakanana paryantam sampurnam birametiah

The top of the crucible should be pierced with an iron rod/needle and then the mould should be picked out from the fire with tongs. The top of the mould should be cleared with a fired stick and then the molten metal should be poured into till the mould is filled up with the molten metal.

8th Stage

Now for breaking the mould and getting the casting:

Sitalatvam ca yatayam pratimayam svabhavatah
Sphotayemmrttikam dagdham biragdhologhuimakah

The mould and cast will cool down slowly. Then the fired mould should be broken to get the metal casting of the image.

9th Stage

The metal channel (created due to pouring action) and the extra portions should be removed from the image:

yatra kkapyaghikam pasyoiccaranoistat prasantayet
nalakam chedayeccapi pascadujjvalatam nayet

The extra portion should be eliminated by cutting off. Then the metal image should be polished by rubbing.

10th Stage

The creation of a new metal image

Tato dravyamayi sacra yatha madananirmita
Jayate tadrsi saksadangopangopasobhita

Thus a complete ornamented, image is created with all parts perfect.[10]

Closer to our context of medieval Bengal we have a number of Sanskrit texts on *rasayana*, which throw some light on the early knowledge of metallurgy and chemical processes. These references are useful for a glimpse of the general knowledge of metal working and applicable to all kinds of product manufacturing, viz., images, utensils, coins and other miscellaneous items. We have collected a few such unpublished manuscripts on *Rasayana* of which one or two have been partially described by late Acharya P.C. Ray. Of these one, *Suvarnatantram* has been transcribed in hand writing and we have begun to analyse some portions. This text appears to be dealing with alchemy. However, it contains important clues to some aspects of metallurgy (medieval times). It reveals the knowledge of eight metals, processes of alloying, killing of metals (oxidisation and removal of impurities through application of heat) etc.[11]

Suvarnatantra or *Svarnatantra* – essentially focusing on alchemy. Indicates knowledge of the eight metals including gold, silver, copper, brass (*pitala*), iron, mercury – working with heat resulting into chemical change – fusion or alloying of metals is clearly indicated. Adding continuously metal into molten metal in an on-heat crucible.

The most important portion which reveals direct knowledge of working with the different metals: the 8 metals in the molten state being treated with a mixture of pure ointment rubbed in oil and kept in oil for 20 days – further treatment with the application of heat is also indicated. The orpiment in the oil the killed and the oil extract when applied to molten copper is said to transform copper into gold.

Similar treatment to tin and bell metal would give rise to silver; copper, iron, brass and silver would yield gold.

The *Ain-i Akbari*[12] also contains significant data about metallurgy and was a universal text in the sense that it was a compilation containing the contemporary processes in vogue. The whole process of minting metal currency was quite old and some of the medieval coins of Bengal Sultans were said to have come out of mints located in Bengal and even Assam.[13] Therefore some of the details mentioned in the *Ain* is relevant for the medieval Bengal context. The chapters on the processes of refining gold and silver[14] opens before us the details of the minting process. Although there is a strong focus on the minting of gold and silver denominators, yet the references to alloying processes like the process of Kukrah, Bugrawati using copper and lead reveal knowledge, perfections and uses of medieval minor metal technology in varied manners. Copper coin minting is also referred to in detail, illuminating a separate dimension of copper works in the medieval times. However, the most interesting portions in this chapter deal with a discussion on the origin of metals and on the specific gravity of metals with intricate measurements and descriptions, mentioned above.

That the *Ain-i Akbari* data is relevant for our project is also clarified when we look at the actual objects like coins, images, utensils found from archaeological sites and historical contexts in Bengal.

Image Crafting

Talking about the actual work of image crafting in Eastern India and historical Bengal especially, one should start with the early medieval data. The early medieval tradition traces the Eastern Indian metal image manufacturing to the different regions of the present states of Bihar, West Bengal, Bangladesh, Nepal and Assam. But instances of images come from various sites located in this part of India. Information from different sources indicates that from Eastern India bronze icons with high artistic excellency were known to have been found from the following major sites. Bronze icons were recovered from Hahara, near Sonapur and Odalbakra Kahalipara (both from dist. Kamrup), and also from Bijulibari, district Darang all in the state of Assam. From Bihar hundreds of icons were discovered from Antichak and Sultanganj (dist. Bhagalpur), Chausa (dist. Bhojpur), Jaipur, Fatehpur and Kurkihar (dist. Gaya), Nalanda (dist. Nalanda), and Sultanganj (dist. Bhagalpur) etc. From Jharkhand the remarkable site is Aluara (near Chandankiyari, dist. Dhanbad) where a few bronze icons were discovered. From Orissa lots of bronze images have been recovered from one of the remarkable sites, Achutajpur. From West Bengal, bronze images are recovered from Pandua and Jagjibanpur (dist. Malda), Sagardighi (dist. Murshidabad) and Gangarampur and Sallas (dist. West Dinajpur) etc. From Bangladesh hundreds of bronze images were recovered. The few important sites are – Paharpur (dist. Naogaon), Jhewari (P.S. Anwara, Dist. Chattagram), Mahasthan (dist. Bogura), Mainamati (dist. Comilla), Raghurampur, Rangpur, Sonarangpur and Paikpara (Vikrampur, dist. Munshiganj). A gold plated bronze image recovered from Balaidhap of Mahasthangarh, has been preserved at Varendra Research Museum. One unique silver image of Basudeva was recovered from village Churain, near Vikrampur, which is preserved in Indian Museum. These icons are well preserved in Indian Museum, Asutosh Museum, Varendra Research Museum in Rajshahi, Bangladesh, State Archaeological Museum, Bangiya Sahitya Parishad, and Mahasthan Museum etc. While Nalanda and Kurkihar were noted centres of manufacture in Bihar, Vikrampur and Jhewari were important centres in the historical Bengal.[15]

1. So far as technology is concerned the early medieval image manufacturers reached a high watermark of excellence. Extreme finesse was required for the modelling and casting processes involved in the making of the intricate pieces of Pala–Sena images. So far as metal content of these images is concerned, we have an assured report given by Riederer, whose observation on the metal composition of *Pala* images is, "Except from an early statuette from the 9th century which is made of a lead-tin-brass, almost all other statuette consist of a very uniform and typical type of a lead – brass with about 10–20 per cent zinc and with 2–10 per cent relatively elevated amounts of lead. As a matter of fact, here too some objects do not fit into this group because of their analytical data, and are seen to be made of pure copper."[16] However, most of the *Pala* metal images fit into a discernible group of craftsmanship and metal components. So far as casting is concerned, there were two processes: solid casting (Ghana) and hollow casting (sushira) for lighter weight images. The first example is referred in the Manasollasa verses mentioned above. The wax model is first prepared and coated with clay very carefully. The mould with the wax is heated, when the wax drains out through an opening provided for it the molten metal is poured through the same and allowed to cool. In the case of hollow cast a model is made of clay or saw dust or some such ingredient and covered with wax. The thickness of the wax determines the thickness of the actual image. Then the model coated in wax is further coated with clay and dried in the sun. Finally the whole piece is heated. The wax melts out through a channel already made in the body of the clay coating. In its place molten metal is poured from a crucible. Molten metal takes the shape of the model inside. The metal is allowed to cool. Finally both the outer clay coating and the inner core model are broken up. The hollow cast image remains to be finished by further grinding and polishing with bobs.

Although the techniques of casting remain almost the same, the medieval images were, quite visibly, of mediocre to crude craftsmanship, compared to the early medieval ones. The reason for this could be many, which is a matter of discussion. However, contrary to the notion that religious image making was completely stalled in the post-Sena rule period, our findings prove that the tradition was alive, and most probably received a boost in the heyday of the Vaisnavite resurgence under Chaitanya. Surveys of museum artifacts reveal the fact that the tradition of making brass and bronze images of religious deities was alive in the medieval period. Images of Buddhist deities like Tara, Buddha in *Dhramackrapravartana Mudra, Jambhala, Nairatma* were found within West Bengal in the context of medieval Bengal and probably belong to the early part of the period. Dual images of *Visnu Lakshmi, Uma Maheswara, Radha Krishna,* single images of *Sita, Jambubana, Sugriva, Radha, Manasa, Mahisasuramardini, Balarama* have been identified in the context of late medieval Bengal. Images of this period come from Bankura and Burdwan, West Dinajpur, Midnapur, Nadia and Bangladesh. The latest variety which probably became popular in the 17th–18th centuries is the image of the *Chaitanya Mahaprabhu* and the whole pantheon of neo-Vaisnavism. Looking at the various effigies and the style of modelling we see a likelihood of tracing a pattern of evolution through the study of these images. On the face of it there seems to have been at least three phases through which the image making tradition continued between the medieval and colonial times.[17]

The colonial records reflect that in Bengal images were principally being made in Nabadwip, Krishnanagar and Kolkata. These images are of Hindu gods and goddesses. Although they are now mostly made of brass, earlier they were made of a combination of 8 alloys, *astadhatu*. This genre represents the more classic of the metalworks in Bengal.

2. We have very illuminating references to metalcraft artisans in the colonial records. The tradition of image crafting was kept alive by private worshippers in the medieval and pre-modern times. An interesting piece of evidence comes from the 1880 work of George C.M. Birdwood,[18] where Birdwood mentions the metals used for image crafting in Benares and Eastern India. He encapsulates the old tradition as observed in the late nineteenth century that must have come down since the medieval times. Birdwood refers to the goldsmiths, brass workers and coppersmiths making little images in brass "alloyed with six other metals, gold, silver, iron, tin, lead, and mercury, making with the copper, and the zinc of the brass, a mixture of eight metals, which is deemed a perfect alloy, and very highly prized." "The gold images of Durga, Lakshmi, Krishna, Radha, and Saraswati kept in private houses and worshipped daily, must not be less than one tola (nearly half an ounce) in weight, and they generally weigh three or four tolas."[19] The case of *astadhatu* has to be resolved through chemical analysis of the relevant artifact. However, the current practice in case of specially ordered images for religious purpose is to pierce a hole on the right chest of the semi-completed image and pour down some mercury. A small piece each of silver and gold are added to the main ingredients of copper and zinc to which sometimes a little tin is added.[20] It is quite probable that the image artisans were actually using panchaloha, as the South Indian artisans of the medieval times did.[21] However,

references to gold plating or washing has not yet been confirmed with hard scientific data, but this is a possibility in the colonial period image crafting as Birdwood refers to it.

So far as processing is concerned we do not have lab-based data as yet on the phases of techniques employed. However, apparently both lost wax and sand casting were in vogue as the details of modelling were getting more and more simple. However, there is a recurrent modelling style that we now find in the images made in Navadvip which begins to date from the 17th–18th centuries, where some amount of standardisation is evident. Not only that, our observation reveals that a number of additional treatments were done like channelling, adding runners, piecemeal casting, brazing and soldering, grooving or engraving in case of sand casting and lastly polishing. Every stage of activity required special tools and the work was totally hand made. In case of the images between 5 to 10 cms in height, there was hardly any requirement for a special furnace. In fact temporary furnaces might have sufficed as we find it even now. What was important however, was the alloying process and furnace for melting the metals was normally bigger than the one for casting. The apparent solidity and surface texture as well as colour of the finished items indicate the ratio of metals used and the finesse of heating. In many cases the surfaces are found to be rough and this may not only be due to long-term exposure or erosion. In many cases the imperfections in modelling were sought to be covered by secondary treatments like additional ornamentation by engraving and soldering appendages. Ethnographic surveys reveal the fact that a few rural artisans are crafting small images for religious purposes more or less following the traditional method. However, some ingredients and ratio or purity of the metals may vary and even degenerate. But the basic change has taken place in modelling. Few rural artisans are able to create such fine models as existed in the early medieval or medieval times. When asked the pretext is their lack of funds and the factor of an unreliable market. The contemporary method of operation is basically the same everywhere, with little variations. We may go into the Kalyanpur village example in East Midnapur first. The process involves making a clay model first. For this the clay itself had to be prepared mixing it with cow dung and jute (instead of strips of fine cloth and rice husk or hay). The core model is then coated with strips of wax, made of a mixture of beeswax, *rajan* (resin), heated in nim oil or *karanja* oil (vegetable oil). The wax is specially layered on to the model and each feature of the core model is sharpened on the wax using fine wooden kathis, small sticks and chisels, quill like tools, so that the final image gets a fine modelling. Now fresh layer of clay is put on the wax-covered model. This coating may be given in several layers to make it thick. Then the whole is dried once more in the sun. A small opening is made in the outer coating of clay now to pour the molten metal. Finally the whole piece is put into the furnace fuelled mainly by cow dung and mixture of charcoal. When the wax melts away the molten metal kept ready in a thick specially made crucible *(muchi)* is poured through this small opening. The whole is allowed to cool. Sometimes the semi-hot piece is quenched in water to quicken the process. At the end of the clay outer coating is broken apart and out comes the finished item in bronze or brass, mostly the latter. I have also recorded a few more examples from Mahisdal in the same district and Navadvip in Nadia district. However, there is serious dearth in designing and modeling expertise.

Dokra

The other most distinct genre comprises the tribal craftsmanship in *dokras*, making images of gods, animals and human figures in a distinct style. Comparatively speaking here we have no dearth of imagination, nor lack of modelling finesse among the contemporary craftsmen. The tradition of *dokra* is now manifest in various different forms in different regions all over West Bengal, Bihar, Orissa, Chattisgarh and Madhya Pradesh. Although the regions of Bengal, Bihar and Orissa in the current scene comprise different states, in the context of pre-colonial and colonial Bengal history, they were a singular entity, geographically as well as politically. Therefore in order to trace the roots of this tradition in history, we have to take in the distinct metal-crafting groups from these three regions into consideration in order to get a holistic picture of the evolution of the metal craft tradition of West Bengal. This is more so because with the passage of time the tribes are found to have moved from region to region within this broad orbit. One such mobile component in the eastern India has come down in history and is presently called the *dokra*. They call themselves *maral*. These people and the traditional Karmakars, who have moved between Jharkhand and West Bengal are the *dokra* artisans today. The tribal components of image making were first mentioned in the colonial records. Older references are not really available on this section. Verrier Elwin had described the tribes of Marrias, Sarras, Gadaba and Bondos using different kinds of metallic ornaments made by themselves, especially of thin brass plates.[22] Hunter's accounts mention the Kamars of Bengal.[23] The Kamars are generally identified as blacksmiths. However, at present the kamars are also engaged in minor, non-ferrous productions. It is mostly the *karmakaras* who form a category that we find handling the craft at present. However, the tribal process of making images by lost wax method has not been quite highlighted before the second half of the last century. Ruth Reeves is one of the first researchers who undertook extensive fieldwork and reported on the Bankura Dokras including a detailed description of their methods of casting. Curiously enough, the traditional style of lost wax is kept alive among the *dokra* workers, numerically so, more than the rural Kansakaras who make traditional religious images. However, the modelling and use of metal composition for the *dokras* are different.

We have good examples of historical *dokra* of the region from the 18th century onwards in the different museums in Kolkata. Our survey of the Gurusaday Museum, Ashutosh Museum and the State Archaeological Museum showed that the oldest collected items date back to 18th century. However, we could discern a changing shift from regular religious imagery to *dokra* style of modelling as well as strip casting in a few older medieval pieces in the store of the State Archaeological Museum, West Bengal. This may date the transition from regular religious image making tribal castings probably sometimes in the 17th century. Or this might have been a result of the extension of the Sanskritisation of some peripheral tribes towards the end of the medieval times, due to which the tribal artisans began to emulate the rural counterpart and provided with new styles of modelling, even new twist to the subjects. The lack of wherewithal forced them to innovate with modelling in strips of metals, so that less metal would be required. Ruth Reeves has thrown considerable light on the individual methods of Dokra workers in the region, like the Mals of Banakura as followed in the middle of the last century.[24] Our report is based on older reports, a scientific project report conducted by the School of Material Science and Technology, Jadavpur University,[25] and our own survey conducted by Banani Bhattacharyya in Bikna, Bankura district.

The report of the School of Material Science and Technology gives the following description of the method involved in *dokra* casting: the *dokra* images are made both as solid as well as hollow castings, although the latter is more common at present. The individual items weigh from a few grams to a few kilograms. Heavier castings of bigger designs are usually made in pieces, which are then joined together by lead-tin soldering. Non-ferrous alloys are used for casting and in most cases they are yellow brass with various compositions of copper and zinc, which contain small amounts of tin, lead etc. The raw material used by *dokra* workers is basically brass scraps, which naturally contain these metals. The wax used by the *dokra* artisans is usually of three types: (a) natural beeswax; (b) a plastic recipe based on beeswax, some amount of paraffin wax, sal dammer resin etc.; and (c) sal dammer resin or *gala*. The local laterite soil is used as clay for the moldings.

Our own survey in Bikna, Bankura district, conducted in 2003–4 shows that the first step is to make a clay core according to the shape of the object. The *dhuna* is a natural resinous material, which has been boiled with mustard oil to produce a hard gleaming black substance. This is held over glowing embers to make it pliable. Flat strips are pulled out from this lump and arranged on a wooden slab. These will cover the entire clay core and later become the solid metal sheath. Finer strips, some twisted, some plaited are added over the smoothened *dhuna* base as garnishing for the design. These make up the extravagant loops and embellishments, which express the artist's cravings. After this the object is given a preliminary coating of mud to keep the sunlight out. This resinous material creates the textures and contours of the dressing. Designs depend on the preferences of the regional people in local myths, social idioms, aesthetic choice/indexes. The thickness of the strips corresponds to the depths of the metal covering. In case of an image, additional features like ears, noses, hands and feet are carved out of wax and remain solid metal, without the support of the clay core. If the design is complicated, it is made up in different parts, cast separately and then attached together with metal pins and wires.

A layer of mud mixed with sand is applied over the object. The craftsman gently moulds the mud over every contour of the design so that the impression gets engraved on the cast. An appropriate spot, depending on the object is selected for placing the channel. This channel is made of bamboo stick. This is removed once the mud is dried and set.

Finally a rough mud covering is slapped over the object. Over the opening of the channel, a funnel is added which becomes the crucible. The metal is now to be used. This comprises of pieces of scrap brass, bought from local metal smiths or recycled from whatever was left over from a previous job. These pieces are kept on the crucible. The Bengal artisans often cover the crucible partially leaving an opening so that one can judge the melting process by observing the metal within. The kiln is made with bricks and a number of moulds covered with cow dung are placed horizontally over hot coals and allowed to bake. The *dhuna* melts quicker so that the crucible is heated first. Later the entire mould is put into the kiln. The time given for firing is generally set according to experience and gut feelings of the artisan. Then each object is removed with long iron tongs. The molten metal is stirred and the pieces are tilted to make the liquid metal settle by putting the mould in a vertical position so that the metal may flow in through the chute and fill the space created by the melted dhuna. When the mould cools down the outer shell is cracked open with a hammer and out comes the artist's creation.

Utensils and Miscellany

The eighteenth century survey records on Bengal, Bihar and Orissa provide population and occupation data. Hunter's accounts mention the Kansaris and the Kamars of Bengal.[26] As already mentioned a section of the tribal Karmakaras have gone into producing *dokra* items. But the Kansaris are age-old workers in brass and bronze/bell metal. These identities were formed in early medieval Bengal and literature of the medieval period makes passing mention of the craft and the artisans. Now many of them retain their old occupation. But this is now mostly restricted to making of utensils in case of bell metal although brass is used for several new shapes.

At the level of history, what is remarkable is the profusion of brass and bell metal utensil production from the medieval times, probably under the influence of Islamic court and elite lifestyle, for a number of new forms of shapes with new uses come into vogue. A survey of the Hazarduary Museum throws light on these new forms that emerged.[27] New forms like *pandan, itr dan, pikdan, masladan, gulabdan, aftaba, octagonal* compartmented trays, *pan dabar* etc., emerged for regular use in a syncretistic society, while special items like *jalidar alam, chawar,* flags for royal processions, brass and bronze fitted weaponry and furniture were part of court life.[28] The work of the Kansaris was cut out for them in the medieval towns of Sonargaon, Pandua, Gaur, and Jessore between 13th and 15th centuries, and 16th–18th century townships of Navadvip, Dhaka, Murshidabad Kasimbazaar, Hooghly, Chattagram, which appear to have flourished with influx of the aristocracy and wealthy inhabitants – having a high standard of life and demand for luxury items.[29] In fact bronze and brass were of everyday, mundane use. The *mangal kavyas, gitika* or ballads of medieval Bengal.[30] throw some interesting light on general lifestyle borne out by other findings. The colonial records really illuminate on the caste Kansari and Hunter's Accounts provide facts and figures which reveal that the Kansaris were producing the usual brass and bell metal/bronze items for all classes: well-to-do and the peasants.[31] A survey of the brass and bronze works in the different districts of contemporary West Bengal has produced useful data. The main items that we found to be manufactured comprise the different varieties of *thala* (dinner plates), *bati* (small bowls), glass, jug, *ghati* (tumbler), pail, tray, *kalsi,* (pitcher) gongs, utensils for religious purpose like small *ghatis*, *ghanta* or bells, *rekabi*, *pradip* or small lamps of different varieties.

So far as technique is concerned it is age-old. From the early days of Chalcolithic civilization, the coppersmiths were employing the technique of cold hammering cutting, bending, grinding and polishing a piece of metal into the required shape and finishing. Next stage is annealing, which requires heating the metal above its recrystallisation temperature, (for copper the temperature required is 500°C) which relieves the strain within the metal and the atomic structure becomes mobile. The metal piece regains shape when it is once more cool and repeated hammering while it is getting cool serves the work of shaping. This process allows the smith more freedom and eases the job of shaping.[32] Since not much of pyrotechnology is involved in annealing, the task is simpler. However, the last method is most effective in case of intricate or difficult objects. Apart from actual preliminary shaping of an object, further processing like thinning, pointing, folding, rimming, lipping, riveting, brazing, lapping, polishing, engraving, finer finishing with chiselling etc., are essential in giving an utensil the required look and shape, tailor-made for different uses. This is followed by hard polishing – primarily done in the old days by rubbing leather or mace stone repeatedly on the surface, now done mechanically by using a rolling grinder pulley to which a cloth bob is attached. Many of these processes were tried and tested by the coppersmiths and brass workers of ancient times. My surveys in the Murshidabad, Nadia, Midnapur and Bankura districts and in the city of Kolkata illumine the different variations of the craft with minor modifications, local styles and innovations, both in case of brass and bell metal items.

3. If we come to the question of metallurgy, the development and proliferation of the brass production is an interesting phenomenon. For brass the requirement was for the zinc metallurgy to develop. Excavations at the most important of all zinc mines in India: the Zawar mine of Rajasthan reveals that it was worked since the beginning of the first millennium. The excavation also revealed a series of structural remains, which were actually furnaces used for distillation of zinc, which requires a temperature above 900°C.[33] It is, therefore only natural for the *kamsa* products, an alloy of copper and zinc, to be a part of metal utensil repertoire of the early South Asian people. However, in most cases now the *kansaris* recycle the old broken utensils to get their raw material, since the colonial times. However, there is this part of the minor metal craft tradition that we are trying to look at: viz., the problem of the source of raw material.

The case of zinc is resolved by the supposition that the ancient miners, extractors and distillers of zinc sold their raw material to the itinerant tribes, who brought it to the eastern Indian smithy. So far as copper is concerned, the Singbhum and Hajaribagh Districts have rich deposits even in outcrops, which were presumably used since early historical times. The intimation about mineral deposits in the region first came in the colonial period from a Mr Jones in 1833. W.W. Hunter records this fact. In 1854 J.C. Haughton published an account of the copper veins in Singbhum district. The present veins of copper are worked near the Bankura and Purulia districts, near the river Subarnarekha.[34] Valentine Ball's report[35] of the copper veins in this area may contain a hint for an old working in the copper veins. Copper mining veins are also located in the stretch between Ghatshila and Kharagpur regions. Chaibasa in the Dhalbhum region, bordering on Midnapore had been identified by Hunter as another area for minerals in the late 19th century.[36] However, the question of availability of tin is a problem. In fact, in their recent findings, N. Kochar, R. Kochar and D.K. Chakrabarti have hypothesized about a possible line of source for tin ore in the Harappan times.[37] There are minor scatters of tin in Gujarat. But the Tusham area in District Bhiwani, Haryana, has a chance of being a major source in view of the region's proximity to the distribution of Harappan sites in Haryana. They report that Tusham's location may also be related to the rich copper bearing zone of south-west Haryana and north-east Rajasthan. More important still is the fact that the tin bearing vein is now found to extend to the Jhunjhunu area in Khetri copper belt. This copper belt also incorporates the adjacent areas of Mahendragarh and Narnaul in Haryana and the Nim Ka Thana belt in Rajasthan. Four belts have also been identified by Dahavalikar, Raval and Chitalwala, as possible sources of Harappan tin and copper: (1) Khetri, (2) Jhunjhunu, (3) Amba mata and Ku-Chitrasani belt extending from Sabarkantha in Gujarat to Sirohi District in Mewari Rajasthan, and (4) Haryana.[38] The possibility of continued exploitation of these old, traditionally known sources for tin in the medieval and colonial times cannot be ruled out. Another possible source area is the north-eastern Himalayas. However, we have to account for the human input in all these importing of metals and think in terms of involvement of different sectors of people in the development of the metal craft in greater Bengal, viz., the itinerant tribal smithy from other regions, the local miners and smelters of copper, probably tribal as well, the brass worker or *kansari* of different specialisation working at different phases of operation from hammering to polishing and engraving.

The Rural *Kansari* or Utensil-maker

Finally, a resume of the traditional artisans in the job:

In West Bengal in the districts of Burdwan, Bankura and Midnapur the metal workers who crafted brass idols and miscellaneous items are called *ure Kamar* or *karmakar*. An interesting point was made by Mukherjee[39] about the term *ure*, which she suggests means flying. Thereby there is an implication of itinerant smithy here. This factor is also evident when we note how some of the tribes like the *Asurs* who were found in the early 19th century in Bengal–Bihar border, were known to have come from Bengal proper. These moving or flying smiths have been seasonal visitors in villages and settlements since early times. Itinerant metal smiths have been identified in very early societies in India since proto-history periods. The origin myths of many of these metal-crafting tribes also point to their mobile occupation. This is sometimes evident in their endogamous names. For example one of the endogamous sub-castes within the *kasera* group in Benares are known as *purbiya*, meaning that they came from the east.[40] The *karmakars* bear surnames like *Mandal* and *Kaviraj*, distinguishing their earlier traditions as village *mukhia* and *Vaidya*, probably. We find that a number of *Rana* artisans have moved into West Bengal from Orissa, *Thateri Ranas* are also a group of brass workers from Uttar Pradesh but at present they are in West Midnapur. The Chandanpur Hat brass craft is mostly manufactured by this community. Midnapur houses the Oriya artisans while Purulia borders on the Chattisgarh component and West Midnapur extends into Jharkhand contingent. The techniques of utensil manufacture is similar although in case of image crafting the alloying components and casting techniques vary from region to region as do stylistic features. In Midnapur a traditional centre of bell metal manufactory was located in the Egra Police station, now shifted to the Potashpur P.S. Another very old centre mentioned by Meera Mukherjee and still continuing is centred in Kharar–Ghatal – all in East Midnapur.[41] The most famous centre mentioned in the colonial records as well as in the work of T.N. Mukharji, published in 1888, was Khagra in Murshidabad[42] (spelt Khankra in Mukharji). Other places in Bengal mentioned were Calcutta city, Kanchannagar in Burdwan, Rajshahi, Islamabad in Dacca, Bansberia in Hugli. The Dhaka artisans of pital utensils (brass) have now settled in Saidabad near Bahrampur in Murshidabad as our survey revealed. The older Khagra contingent in the district is almost closed down, baring a few households who are struggling on with their traditional occupation. In fact the changes noted from the report provided in the Herculean work of Meera Mukherjee in 1970s during our on-going survey in the new millennium, are so significant that the task has become a social challenge to the small team that we have working on the current INSA sponsored project.

Acknowledgment

I would like to end this article with thanks to the following persons and institutions without whom I could not have progressed:

First my teammates:

Dr Pranab Chattopadhyay at Centre for Archaeological Studies and Training, Eastern India and Smt. Banani Bhattacharyya, my Research Assistant.

Dr Gautam Sengupta, Director, Directorate of Archaeology, West Bengal and Secretary, Centre for Archaeological Studies and Training, Eastern India for extending valuable support.

Dr Bimal Bandopadhyay, Superintendent, Calcutta Circle, Archaeological Survey of India.

Dr Shakti Kali Basu and the officers of the Indian Museum, Kolkata.

Sri Soumitra Sarkar, Officer-in-charge, Ashutosh Museum.

Sri Pratip Mukherjee, Curator, State Archaeological Museum.

Sri Tapas Bandopadhyaya, State Archaeological Museum.

Sri Ashish Chakrabarty, Curator, Gurusaday Museum.

Sri Sadhan Chandra Deb, Curator, Malda Museum.

Notes

1. *Manasara*. Ch. LXVIII, *Madhuchchhista-vidhana*, in P.K. Acharya (ed. and trans.), *Manasara*. Series No. 2, *Indian Architecture according to Manasara-silpasastra*. A Complete and Unabridged Version, Delhi, 1998 reprint, p. 86.
2. P.K. Acharya (ed. and trans.), *Manasara*. Series No. 2, *Indian Architecture according to Manasara-silpasastra*. A Complete and Unabridged Version, Delhi, 1998 reprint.
3. Ruth Reeves, *Cire Perdue Casting in India*. New Delhi: Crafts Museum, 1962.
4. M.K. Krishnan, *Cire Perdue Casting in India*. New Delhi: Kanak Publications, 1976.
5. *Karanagama*. II, V, p. 41.
6. *Suprabhedagama*. XXXIV, p. 21.
7. Visnu Samhita, *Patala*. p. 14.

8. *Sukranitisara*. IV, 4, p. 72.
9. Ibid., pp. 152–53.
10. T. Shama Shastry (ed.), *Abhilashitarth-Cintamani*. Government Oriental Series, University of Mysore: Oriental Library, *Journal of Indian Society of Oriental Art* IV, No. 2, Dec. 1936, pp. 142–44; a reference is available also in Ruth Reeves, op. cit., p. 33.
11. Praphulla Chandra Ray, *History of Hindu Chemistry from the Earliest Times to the Middle of the Sixteenth Century*. Vol. II, London: William and Norgate, 1909, New Centenary edition, Kolkata, 2002, pp. 28–29.
12. *Ain* 13 and 14 in *Ain-i Akbari* by Abul Fazl, The Complete English Translation by H. Blochmann, in 3 volumes published by the Asiatic Society, reprint edition, Kolkata, 1993, pp. 40–45.
13. Robert S. Wicks, "Fiscal and Monetary Policy in Southeastern Bengal, Assam and Arakan". *Money, Markets and Trade in Early Southeast Asia: The Development of Indigenous Monetary Systems to AD 1400*. New York: Cornell University, Chapter 3, 1992, pp. 66–109; A. Karim, *Ciorpus of the Muslim Coins of Bengal*. Dacca: Asiatic Society of Pakistan, 1960, pp. 158–59.
14. *Ain* 7, 8 and 9, in *Ain-i Akbari* by Abul Fazl, The Complete English Translation by H. Blochmann, reprint edition, Kolkata, 1993, translation pages, pp. 23–27.
15. Mentioned in Debala Mitra, *Bronzes from Bangladesh: A Study of Buddhist Images from District Chittagong*. Delhi: Agam Kala Prakashan, 1982; R. Basak and D.C. Bhattacharya, "Catalogue of the Archaeological Relics in the Museum of the Varendra Research Society", *Rajshahi*, 1919; Stella Kramrisch, "Pala and Sena Sculpture", Rupam No. 40, 1929; K.P. Jayasawal, "Metal Images of Kurkihar Monastery", *Journal of the Indian Society of Oriental Art*, Vol. II, No. 2, 1934; S.K. Mitra (ed.), *East Indian Bronzes*. Calcutta: Calcutta University Press, 1979; Susan L. Huntington, *The Art of Ancient India: Buddhist, Hindu, Jain* with contributions by John C. Huntington, New York and Tokyo: Weatherhill, 1985, pp. 387–413; Bimal Bandopadhyay, *Survey of Indian Metal Sculpture*. Delhi: Sudeep Prakashan, 1987, pp. 15–31.
16. J. Reiderer, "The Relation between the Composition of North Indian Statuettes of Copper Alloys and the Region and Date of their Origin", in M.C. Ganorkar and N. Rama Rao (eds.), *Role of Chemistry in Indian Archaeology*. Hyderabad: Birla Institute for Scientific Research, 1999, pp. 98–99.
17. Catalogues of objects from the State Archaeological Museum and the Ashutosh Museum, Kolkata.
18. George C.M. Birdwood, *The Arts of India*. Great Britain: British Book Company, 1880.
19. Ibid., p. 157.
20. Observation from my survey of the Navadvip workshops.

21. Personal information from P. Gayatri, Department of Archaeology & Museums, Hyderabad and Sharda Srinivasan, IISC, Bangalore.
22. Verrier Elwin, *The Tribal Art of Middle India: A Personal Record*. London: Oxford University Press, 1951 reprint, pp. 11–14 and p. 55.
23. W.W. Hunter, *A Statistical Account of Bengal*. London: Trubner & Co., 1877.
24. Ruth Reeves, op. cit., pp. 36–54.
25. Progress Report on Documentation Project: Quality Assessment and System Improvement of Tribal Metal Castings, submitted by the School of Material Science and Technology, Jadavpur University, 1992, DST project.
26. W.W. Hunter, op. cit.
27. Courtesy, the ASI, Calcutta Circle.
28. This matter is also discussed by Amita Ray, "The Bell and Brass Metal Craft of West Bengal: A Dimension of Indian Art", *Pupul Jayakar*, 70 Volume, New Delhi, 1986, pp. 359–73.
29. A number of works throw light on the urban centres of medieval India, viz., Imtiaz Ahmad, 'Urban Centres in Medieval India', in V.K. Thakur (ed.), *Towns in Premodern India*. Patna: Janaki Prakashan, 1994, pp. 217–24; C.A. Bayly, *Rulers, Townsmen and Bazars*. Cambridge: Cambridge University Press, 1988; S. Subramaniam (ed.), *Merchants, Markets and State in Early Modern India*. New Delhi: Oxford University Press, 1990; G.R.G. Hambly, "Towns and Cities: Mughal India", in T. Roychoudhury and I. Habib (eds.), *The Cambridge Economic History of India*. Vol. I, Delhi: S. Chand, 1984, pp. 434–51 and other works on local history.
30. Chandi Kavikankan, Srikumar Chattopadhyay and Viswapati Chaudhury (eds.), Calcutta University, 2002 reprint; Kavi Jagajjiban composed, *Manasamangal*. Ashutosh Das and Surendrachandra Bhattacharyya (eds.), Calcutta University, 1984 reprint; Dineschandra Sen compiled and edited. *The Ballads of Bengal*. Vols. I, II and III, Delhi: Motilal Banarasidass, 1988 reprint.
31. W.W. Hunter, op. cit., Delhi: D.K. Printworld, 1973 reprint, mention this while describing the material condition of the people in each district.
32. Discussed in Arun Biswas, *Minerals and Metals in in Ancient India*. 2 Volumes, New Delhi: D.K. Printworld, 1997 and Nupur Dasgupta, *The Dawn of Technology in Indian Protohistory*. Calcutta: Punthi Pustak, 1997, pp. 392–96, which is modelled on various scientific reports and H.D. Sankalia, *Some Aspects of Prehistoric Technology in India*. New Delhi: Indian National Science Academy, 1970.
33. Syed Jafar Mahmud, *Metal Technology in Medieval India*. Delhi: Daya Publishing House, 1988, pp. 77–84.
34. W.W. Hunter, op. cit., Vol. XVII, pp. 35–38.

35. Valentine Ball, *A Manual of the Geology of India*. Part III, Calcutta: Geological Survey of India, 1881.
36. W.W. Hunter, op. cit., p. 99.
37. N. Kochar, R. Kochar and D.K. Chakrabarti, "A New Source of Primary Tin Ore in the Indus Civilization", *South Asian Studies*, No. 15, 1999, pp. 116–18. The case of tin bronze in eastern India is discussed to some extent by P.K. Chattopadhyay, *Archaeometallurgy in India: Studies on Technoculture in Early Copper and Iron Ages in Bihar, Jharkhand and West Bengal*. Patna: K.P. Jayaswal Research Institute, 2004, pp. 72–80.
38. M.K. Dhavalikar, M.R. Raval and Y.M. Chitalwala, *Kuntasi: A Harappan Emporium on the West Coast*. Pune: Deccan College, 1996, pp. 211–12.
39. M. Mukherjee, *Metalcraftsmen of India*. Memoir No. 44, Calcutta: Anthropological Survey of India, 1978, p. 12.
40. W. Crooke, 1896, *The Tribes and Castes of North-Western India*. Vol. III, Delhi: Cosmo Publications, reprint 1974, p. 168.
41. M. Mukherjee, op. cit., 1976, p. 11 and 91.
42. T.N. Mukharji, *Art Manufacturers of India*, specially compiled for the Glasgow International Exhibition, Calcutta: Art Manufacturers of India, 1988, pp. 191–92.

8 Craft Technologies and Craft Communities in Peninsular India

An Overview of Sixteenth to Eighteenth Centuries

Vijaya Ramaswamy

The period from sixteenth to eighteenth centuries, especially the eighteenth, are very important in peninsular India from the point of view of the economic historian. A study of this crucial phase of South Indian history has given rise to a number of theories and an equal number of questions. To those historians of the seventeenth and eighteenth centuries who had viewed the history of these times through the prism of the Mughal empire and its downfall, these centuries constituted crisis years in Indian history. One of the earliest examples of this historiographical position is an essay by Herman Goetz titled "The Crisis of Indian Civilization in the Eighteenth Century" published in the 1930s.[1] The notion of crisis especially in the agrarian sector with a decline in crafts as its scatter effect, began with Irfan Habib's theory of the agrarian crisis in the nineteen sixties. Barun De's Presidential Address to the Indian History Congress in its 49th Session at Dharwad, focused in fact on the 'Crisis' School of eighteenth century India.[2]

As scholars have made a paradigm shift from a Delhi centric approach to a more balanced perspective of the regional economies, some of them have looked at these centuries as the era of regional prosperity under Deccani states. Frank Perlin used the term 'Proto-Industrialization in the Periphery'[3] to describe the economic landscape of the region. Crafts and commerce began to prosper under the Nayaka kingdoms of Madurai, Tanjavur and Ikkeri, the Nizami State of Hyderabad and the princely state of Mysore under the Odaiyars. The counterpoint to the eighteenth century as the era of decline is the collection of articles edited by Richard Barnett which seeks to look at the vibrant polities of regions As he outlined his agenda, it was time one moved 'beyond the gloomy paradigm of decline' and savoured the richness and vivacity of the eighteenth century.[4]

In recent years monographs, edited books and articles dealing with the historical and historiographical rigmarole that constituted the eco-landscape between the sixteenth and eighteenth centuries, have multiplied. The Oxford University Press deals with this century in its recent publication[5] under its significant series called 'Debates in Indian History and Society'.

The political and socio-economic milieu of the pre-colonial era provides the context for my overview perspective of technology, society and traditional crafts in peninsular India. In this article I shall focus essentially on the Vishwakarma craftsmen although I may draw parallels and comparisons from other crafts communities. While I shall draw primarily upon the lives and crafts of the Vishwakarma community I shall problematise certain specific core issues in the study of craft technologies.

The empire of Vijayanagar witnessed the growing prosperity of certain categories of craftsmen like weavers and smiths. This was due to a combination of factors such as the emergence of a new class of consumers represented by the nobility and officialdom of the state. The Vijayanagar state played an active role in encouraging both production and commerce. The Deccani Muslim states brought new styles into fashion such as doublets, embroidered jackets and turbans. Textile varieties like *mashroo* and *himroo* owed their existence to the Muslim influence. These varieties have an interesting origin. Islam forbids the use of silk by its adherents. Yet the fashion conscious nobility in the Deccan was keen on using silk in its prayer mats. The result of their peculiar requirement was the unique weaving technology of *mashroo* cloth which was a silk cotton mix with the cotton threads on top of the mat, thereby technically fulfilling the Islamic injunction while effectively violating it. *Mashroo* and *himroo* went on to become fashionware not only in the Deccan but also in the Mughal empire and found a lucrative export market in West Asia.[6]

The ubiquitous presence of the European companies – primarily the English, the Dutch and the French – foregrounds technology, crafts, commerce and culture in peninsular India. Craftsmen like the weavers and painters became linked to the larger commercial enterprise of the imperial powers in which Indian merchants became comprador and crafts and commerce moved away from indigenous production and trade centres into the European 'Black Towns'. The regional kingdoms of the Nayaks and princely states provided competing focal points of power and patronage. The growing demand for craft products in the various regional principalities and the centrality of craftspersons to the economy led to many scholars perceiving these years as a boom period followed by a bust in the late eighteenth century. Competition leading to growth in production gradually petered out as Britain established its supremacy over its rivals and emerged as overlord with monopoly over the crafts and commerce of its Indian colony.[7]

V.I. Pavlov, the Russian scholar, formulated the theory of uneven development of craft and industrial production in the pre-colonial era. In his book *Historical Premises For India's Transition to Capitalism*[8], he speaks of the pre-colonial Indian economy as a "multi-sectoral" one in which certain sectors such as the textile industry constituted nascent capitalist pockets in which craft techniques and technology was way ahead of many other craft sectors. In fact for the textile industry in South India one could speak of proto-industrialisation as a possible stage of capitalist development as I have tried to show in one of my early articles.[9] To a lesser extent certain specialised metal crafts benefited from the growing demand both from the domestic elite and the export market especially West Asia. Bidri ware and diamond production in the Deccani States especially Bijapur and Golconda, would be the best examples of the crafts benefiting from the commercial boom.

Traditional Technologies and Origin Myths of Craft Communities

Weavers believed that every part of the loom represented a particular aspect of divinity. A devotional folk song of the weavers which can be dated roughly to the late fifteenth to early sixteenth century (on the basis of the reference to Vira Vasantha Raya, a Vijayanagar feudatory) illustrates this point. The composer was Ramasami Kavirayar who is usually dated to the fifteenth century. I quote:

> Turning into yarn the thread from the lotus[10]
> Devas stretch the threads, mark the picks
> tie to together the threads.[11]
> Adisheshan (the five hooded snake on which Lord Vishnu reclines)
> the strong rope binding the pegs,
> Brahma as the plank
> The threads stretched out and held in place by stones that act as weights

The sticks with holes through which the threads pass
The warp and the woof are
Siva and Parvati[12]

Even the pirn in the shuttle and the foot treadles consisting of two sticks, (in later looms, the treadles could even be four or six) are said to represent the planetary gods like Venus (Shukra), mercury (Budhan) etc.

The primary identity of the Vishwakarma craftspersons stemmed from their community rather than caste identity. The majority of medieval inscriptions emphasise that the artisan belonged to the Vishwakarma *kula* rather than detailing his caste affiliation. The Vishwakarma smiths consisted of five affiliated groups of craftsmen – goldsmith *(tattan)*, brass smith *(Kannan)*, blacksmith (*karuman* or *kollan*), carpenter *(tachchan)* and mason (*silpi* or *kal-tachchan*). Despite belonging to different professions and different *jatis,* these craftsmen claimed a collective entity as *Kammalar* in the Tamil country *Panchanamuvaru* in the Andhra country and *Panchala* in Karnataka. However in many of the medieval inscriptions they are also collectively referred to as the *Rathakarar* or the *Kammala-Rathakarar*. Today they prefer to use the nomenclature 'Vishwakarma' the divine architect from whom they claim descent.

The smiths have a common mythological origin, claiming their descent from Vishwakarma. Two important inscriptions from Tiruvarur (Tanjavur district)[13] and Uyyakondan Tirumalai[14] (Tiruchchi district) or the twelfth century provide us however with a detailed analysis of their social origins.

It was usually in the context of their role as engravers and sculptors that the names of craftsmen at times accompanied by the panegyrical note on their origins and skills occur in inscriptional records. An inscriptions from Macherla[15] in Palnad Taluq in Guntur district dated AD 1111 states that the smiths were descended from Vishwakarma who was the son of Brahma. He is also said to be the father in law of the sun, converting the rays emanating from the sun into divine weapons like the discus of Vishnu. In terms of understanding the history of craft technology this inscription is important because it lists a long geneology of craftsmen and says that they had a proper understanding of their craft in terms of "grasping the 'kshethra' (parameters?) involved in the study of Vastu Shastras". They were therefore successful in cutting of *'linga'*, preparing divine icons, building four kinds of mansions *(prasada)* etc. One of the longest panegyric on the *Kammalar* comes from Alangudi in Tanjavur. Found in the Apatsahayesvara temple the inscription[16] is dated AD 1264 and registers the constructions of a *mandapam* by the *Kammala-Rathakarar* community and the grant of certain endowments to the temple for its unkeep and maintenance. Interestingly, they describe themselves as "the four kinds of Rathakaras of the six kinds of profession". Since the workers in the mint have been referred to separately quite often (the *kammatta*), one can presume that this was the sixth category of craftsmen.

While the mythological origins of the *Kammalar* are found in the form of *prashastis* or panegyric usually in those records where the *Kammalar* feature as temple donors, inscriptions of a general nature sometimes extol specific craftsmen. A record of the period of Rajendra Chola I dated AD 1018 from Tiruvalangadu[17] (also called Palayanur in Chingleput district written in Tamil and Sanskrit contains an order in Tamil converting the village from a *Brahmadeya* (Brahmin Village) to a *vellan Vagai* (non-Brahmin) village. The panegyric in Sanskrit states that the four skilled sculptors (*shilpa kovidaihi*) – Aravamurta, Ranga, Damodhara and Purushottama, the sons of Krishna, well versed in the various forms of mechanical arts, who had their birth in the great city of Kanchipuram, who were clever and wise and born in the Ovi family, engraved the above inscription. One of the earliest eulogical references to craftsmen is found in two inscriptions from Repalli in Tumkur (Mysore) and Ayyangaripalem in Guntur[18] which refer to the architect of the temple of Jalpesa, called Maindarama as "kalgarabharana acharya"[19] (meaning an ornament to sculptors) of the fourth (i.e., sudra) caste and the jewel of the *Vishwakarma kula*.

In terms of their social and ritual status, the origin of the *Kammala-Rathakarar* was defined as being that of a mixed caste, primarily anuloma, i.e., born of high-born father and low-born mother. In the Tiruvarur inscription of the twelfth century from Pandyakulantaka Chaturvedi mangalam[20] (Tanjavur district) and another record from Rajasraya Chaturvedimangalam from Uyyakondan Tirumalai[21] (Tiruchchirapalli district) of the same period, the Brahmin determined the status, rights and duties of the *Kammala-Rathakarar*. Based on the Dharma Shastras and the Smritis of Yajnavalkya, Gautama, Maskara and Bhima, they arrived at the consensus that the *Rathakara* were primarily *'anuloma'* being born of *Mahisha* male *(Vaisya)* and *Karana (Sudra)* female. According to some texts as *'anuloma'* they could perform the *upanayanam* (sacred thread ceremony) and *'sandhyavandanam'* (the special worship of the sun at dawn, dusk and high-noon) but silently (*tushnim* eva). Yajnavalkya however forbids them to read all sacred texts relating to architecture and sculpture like Agastya Vastu Shastra and Visvakarmiya, though they had the right to build temples and sculpt the images of gods.

The consolidation of these early origin myths of the Vishwakarma community was done sometime in the seventeenth, eighteenth and nineteenth centuries. The *Visvakarma Puranam* brought together the various oral traditions of the craftsmen. Although the manuscript is undated one can roughly date it to the colonial period, perhaps mid-seventeenth or eighteenth century on the basis of statements such as the lament of the author that at one time the Vishwarkarma conjured up things through their creative imagination and now they were compelled to work according to company masters.

The *Visvakarma Puranam* says that Vishwakarma wrote the *Mayanool* which is the science of architecture. This contains the details of indigenous hand measurement systems (*Kadam*, ma, *Yojanai* etc.) and refers to the importance of mathematical and astrological calculations in the construction of buildings.

One of the most interesting myths from the point of the interlacing of technology and culture is the origin of the five professions of the Vishwakarma community. The *Visvakarma Puranam* says that Brahma and Vishwakarma together created the universe. In their own special version of the 'Big-bang theory' the artisans claim that the five natural elements formed an enormous egg which burst forth like thunder and the universe came into being. Siva and Vishnu emerged from the blue space and created Vishwakarma and Brahma respectively. Vishwakarma had five faces representing the three smiths (in fact according to the text, of the colour of their faces which were also gold, copper and black) as also the mason and carpenter. Vishwakarma then made tongs out of the power emanating from Brahma and Vishnu and joined it with the nail called Rudram or Siva. He called this *'Kuradu'* which is of course an important tool of the smiths.

This original myth also makes the points that Vishwakarma was born wearing the sacred thread "similar to what the Brahmins wear around their shoulder" this loaded statement making it obvious that their one major purpose was to claim Brahamanical status. The Vishwakarma artisans I have interviewed told me that even now *Avani Avittam* or thread changing function is among their most important ceremonies.

Caste and Crafts

The craftsmen in pre-colonial India were organised in corporations. Appadorai has used the term 'caste guilds' to describe such corporations.[22] The nature and scope of these organisations raise a number of ticklish issues. It would be a truism to state that crafts in India was inextricably linked with the *jati* or caste system. Karl Marx argued that the influence of the caste system on crafts resulted in the confinement of a craft within a small group. The division of labour, even though it existed, was of several generations of craftsmen living at one time and working together at the manufacture of a given article, the trades being hereditary. As a result, "they either petrify into castes or ossify into guilds."[23] Marx gives the example of the Dacca Muslins which he believed were produced "without capital, machinery, division of labor or any of those means which gives such facilities to the manufacturing interests of Europe." He concludes, "It is only the special skill accumulated from generation to generation and transmitted from father to son that gives to the Hindu as it does to the spider, this proficiency."[24]

Within the historiographical tradition of Indian economic history the inherent potential of the craft sector was examined by Irfan Habib in his article, "Potentialities of Change in the Economy of Mughal India".[25] He attributed the relatively slow pace of technological change to the presence of cheap but skilled craft labour in India. However, the system of working within craft guilds has been interpreted by K.N. Chaudhuri as a "high quality trap".[26] To quote him, "Specialisation (in textile technology in India) had gone so far that it would have required very great incentive to induce such men to change their production methods and habits drastically." [27] Given the level of excellence of the craftsmen and the resulting demand for Indian products especially among the elite both at home and abroad, there was little incentive to innovate, a situation which was quite the opposite of what pertained in England in the pre-industrial era. To illustrate, the cotton revolution in England in the eighteenth century because necessary only because of the severe competition posed by Indian textiles, especially calico, which was flooding the European markets including England. Economic compulsions such as high labour costs (in contrast to cheap Indian labour) led to technological innovations and the new production techniques.[28]

To perceive the craft guilds as stultified within the caste system is to ignore the dynamics of South Indian society and economy. Firstly the crafts which formed a corporate collectively known as the *Kammalar* or *Panchalar* did not constitute a single caste or a single profession. The caste of the goldsmith was different from the caste of the carpenter and yet they came together under the generic term 'Vishwakarma'. The professions and role functions of the constituent categories of the Vishwakarma – smiths, carpenters and masons/sculptors also varied widely. Yet they had a strong community identity as 'the group of five'. Further caste in the economy of pre-colonial India did not necessarily play a negative "petrifying role". In the course of the seventeenth and eighteenth centuries specialisations proliferated mainly in response to the western markets being reflected interestingly in the emergence of new castes. *Caingaloon* and *Mambaloon* weavers who are mentioned in the seventeenth–eighteenth century factory records constitute specialist categories. For instance, the *Caingaloon* weavers were those who wove cloth destined for the Malay Archipelago: the word 'kain' in Malay means cloth and 'gulong' means rolled. The supply of gold thread to the *Caingaloon* weavers at Armagaon is referred to in an English East India Company letter dated 1629.[29] Similarly the *Mambaloom* 'painters' were weavers specialising in the production of painted chintz.[30] The Mooree weavers produced superior cotton cloth used as a base for chintz making.[31] The Pattunulkaran or Saurashtra weavers who became specialists in the tie and dye technique, were concentrated in Madurai and rose to prominence in the course of the seventeenth century. Therefore one can conclude that the Indian craftsmen had a dialectic relationship with caste, bending and stretching the caste structure to accommodate new techniques and technologies.[32]

Technology, Crafts and Traditional Beliefs

One of the central questions regarding the evolution of crafts is their link with technology. The emergence of new artisanal *jatis* or castes shows that it was in the nature of craftsmen to form new caste groups on the basis of adaption of new craft techniques. However, in the Indian context one also observes that though craftsmen in the course of the eighteenth and nineteenth centuries were exposed to new and perhaps more efficient technologies introduced by the European companies, they often showed a strong reluctance to adopt them, preferring the existing practices. Is this clinging to tradition a sign of reluctance to imbibe new ideas or is it a matter of appropriate technology?

A prime illustration would be the issue of the effort by the agents of the English East India Company to introduce the coulter both in the Malabar and the Bengal countryside. The iron tipped plough called the *kozhu* was known to Indian farmers for a long, long time and in fact this term occurs in the Sangam text *Perumpanatruppadai*. However it was used judiciously depending on the needs of a particular kind of soil and its suitability for farming. For instance light ploughs were used in fertile areas like the Kaveri basin while heavy ploughs drawn by four or five bullocks were used in the Punjai or dry lands as in parts of Tondaimandalam or Kongumandalam. Madhava's text *Parasara Madhaviyam* refers to the use of a ploughs drawn by six to eight bullocks.[33] The iron tipped plough was used only in these areas which required intensive ploughing. The *Parasara Madhaviyam,* a fifteenth century text attributed to Madhava, the minister of the Vijayanagara king Krishnadeva Raya, makes a curious reference to the use of this device:

> *loha sahitena langala mukhena, praninaha*
> *chinnabhinnam bhavantiti matsyavadat papadhikyam uktam*[34]
> A rough translation of this verse would be:
> with the use of an iron tipped plough (by the farmer)
> more creatures are killed and greater sin incurred than the killing of (hundreds of) fish.

Here what appears to be a moral injunction is in fact backed by strong economic reasons. The purpose seems to have been to prevent the erosion of the substances present in the surface soil like alkaline substances and minerals that are beneficent for crops. This would only have resulted in reduced productivity since inferior soil would have come up. It has been noticed that in the deltaic regions of South India that the seed yield ration was as high as 1:12 or 1:20. Hamilton Buchanan writing about ragi cultivation in the Seringapatam region in the eighteenth century estimated the average seed: yield ratio to be as much as 1:40.[35] This is in striking contrast to the European figures for the same period which range between 1:4 to 1:6.

In 1820, Alexander Walker, made this significant observation about the Malabar plough:

"In a climate where the productive powers (of the soil) are so great, it is only necessary to put the seed a little way into the ground … It must be strong proof that the Indian plough is not ill adapted for its purpose, when we see arising out of the furrows it cuts, the most abundant and luxurious crops. What can be desired more than this?" The labour and expense beyond this point must be superfluous. The Indian peasantry is commonly well enough informed as to his interest and he is generally intelligent and reflecting. He is attached to his own modes because they are easy and useful … ."[36] Dr Wallick, a superintendent of the English East India Company commented on the attempt to introduce the coulter in the Bengal countryside. Speaking before the committee of the English Commons in 1832 he remarked:

> The Bengal husbandry although in many respects extremely simple and primeval in its mode, yet is not quite so low as people suppose it to be … I have known, for instance, of European iron ploughs (the reference here is to the coulter) being introduced into

> Bengal with a view to superseding the extremely tedious and superficial turning of the ground by a common Bengali plough. But what has been the result? That the soil, which is extremely beneficent, has generally received the admixture of the undersoil which has deteriorated it very much.[37]

In the case of the coulter at least it is clear that the rejection of the new technological device was not on grounds of lack of initiative on the part of the peasantry but an awareness of what technology would actually be appropriate for this specific Indian situation. In dry areas or the rain fed punjai lands, the coulter was definitely in use. The awareness of varying techniques and technological usages is reflected in two late medieval texts – the *Tiruvarur Pallu of Jnanaprakasanar* (primarily an account of the community of agricultural labourers called Pallar) and the *Mezhivilakkam,* (literally 'an exposition on the plough') which is of unknown authorship.

Another example of appropriate technology can be cited from the textile industry which was one of the strongest traditional industries in the pre-colonial Deccan accounting for a major share of India's exports.[38] The illustration pertains to the attempt by the European Agents to use the reeling and fulling machines in South India. Writing in the nineteenth century, Elijah Hoole[39] remarked on the reluctance of the indigenous textile industry to use the speedier and more efficient reeling machine. This consisted of iron pegs fixed on a frame with slides, around which pegs the cotton yarn is formed into hanks. Interestingly, Hoole himself offers an explanation for this that the rejection of the device was due to the extreme moisture in the atmosphere which tended to rust and corrode everything made of iron in which case the cotton would adhere to the rollers and the whole yarn would be ruined. The fulling machines were introduced in around 1607. In this device two heavy wooden hammers are lifted by cams on the shaft turned by the water wheel. As the cam passes the head of the hammer, the hammer would get released and deliver a blow to the cloth in the tank. The reasons for the rejection of this device does not seem to have a specific logic unlike in the case of the reeling machine.

One of the oral traditions highly popular among the Vishwakarma community is the story of the magnetic fort built by them which was impregnable. The story finds a detailed description in the *Visvakarma Puranam.*[40] One common version of this oral tradition is that the five artisans of the Visvakarma Kula were at one time entirely in the service of the gods. Since they were celebrities, they possessed the power of creating things by merely imagining them. Their fort made of lodestone was invincible. The location of this fort is said to be on the shore of Ilangapuri (Sri Lanka). As a result of its magnetic powers, the weapons thrown by enemies struck to the fort walls and could not penetrate within. The principal enemy of the Vishwakarma was a king called Karunakaran, who is said to have derived his authority from the Chola Raja. He planted several beautiful women (in some versions they are Brahmin women) in the fort who married the Vishwakarma, thus destroying their spiritual powers and revealed the secret of the fort to the enemy king. This was the use of a particular kind of grass, which was poisonous and could burn up the fort. The fort was blown up and the Vishwakarma were scattered in all directions and forced to work for ordinary mortals, making weapons, ploughs, houses, temples to small things like pins and needles.

Certain points of this story are extremely significant to the economic historian or the craft specialists: (a) The reference to the lodestone fort. Hamilton Buchanan, for instance, has mentioned the finding of laterite and magnetite, i.e., the lodestone on iron mining sites. Besides the awareness of its magnetic properties, the lodestone also may have been used in the process of smelting iron. A Dutch Scholar, Jan Browner, in his recent survey records the statement of Karnataka smiths that they use lodestone in their smelting furnaces since its melting point is higher than that or iron ore. (b) The second point is the use of grass which is described as 'Haraka' in the Kannada version and ' Varagu' in the Tamil version. This has been identified as a variety of wild millet which can be highly poisonous if not properly processed. After processing it was eaten in the form of pancakes.[41]

The final illustration in this section comes from iron mining in the eighteenth century in the Karnataka region. Hamilton Buchanan describes the problem of waterclogging which nagged the miners often resulting in the loss of lives. He describes the existing technology as "crude and unthrifty".[42] After reaching a depth of little more than six feet, the would become waterlogged and hazardous. Very often the mines just caved in resulting in the death of miners. Mining would be eventually abandoned because they failed to find a system by which the water could be drained out and deep drilling could be done.

The poor miners found their own solution to this technological problem. Before entering the mines, elaborate worship was offered to a piece of mined rock or iron ore called 'Karikallu Devaru'. This was followed by the worship of Guddada Amma, literally 'The Mother of the Hills'. The miners also had a special protective spirit, the bhuta figure Muttarayar. Thus the emergence of mining deities in the Karnataka region had as its root cause a technological problem in mining.

Gender Issues in Pre-colonial Crafts

An overarching patriarchal framework can clearly be perceived in pre-colonial craft structures. Women by and large functioned in an ancillary or subordinate capacity in the work of the smiths. In the craft of the blacksmiths, the women are allowed to work the bellows while they are not permitted to forge the iron. In jewellery-making among the goldsmiths and the silversmiths women are engaged in the tasks of polishing, fine cutting and embellishing the designs but not given a primary role in the designing or making of the ornaments. However, among the Gudigara caste in Karnataka, women are said to have taken over the family profession in the case of death of the male member and this included goldsmithy. The goldsmith community called 'Bayala Akkasaliga', literally 'itinerant goldsmiths' seems to have trained both men and women in the art of goldsmithy.[43] One does not hear of women sculptors at all especially in the craft of icon making. In the context of the wheel based industry like pottery, much of the activity seems to have been joint, with both men and women being involved in kneading of the clay and the firing of the pots. However it was exclusively the work of men to turn the pot on the wheel. The buffing or polishing of the pot was done mostly by women.

Among handicrafts, women seem to have played an important role in the handloom sector. Spinning in South India was done, and is still done, entirely by women while in the north, especially in the hills, men can also be seen spinning on a spindle. While the spinning of thread was also done by housewives, it was by and large the sole occupation of widows and unmarried or destitute women. Such female spinners are referred to in the Sangam literature as *'parutti pendugal'* literally 'cotton women' or *'kaimai pendir'* literally 'widowed women'.[44] Medieval craft history is also replete with references to women spinners. Within the Virasaivite (also known as the Lingayat movement), a number of craft persons are referred to in the medieval work *Shunya Sampadane.* The Shivasharana Jedara Dasimayya belonged to the weaver caste and it is said that his saintly wife Duggali spun the thread that he wove into cloth.[45] Remmavve and Kadire Kayakade Kalavve, both have the prefix 'kadire' to indicate their professional status as spinners.[46] A reference to women spinners during the Vijayanagar period comes from the poet Srinatha in his *Palnattu Viracharitra* where he says that in Palnad while the farmers ploughed the field, their wives spun thread.[47]

Women were however excluded from the craft of weaving for ritual reasons. For South Indian craftsmen the loom represented divinity itself and women could therefore not touch an object that was divine and therefore pure because of their periods of impurity.

The twin crafts of both mat weaving and basket-making was largely in the hands of women. In fact the raw material for both, a grass called *korai* was picked only by women. A folk song of the colonial period records the lament by a pregnant *korai* picker:

> As I come to cut the korai growing around my house, my Lord,
> growing around my house,
> you threaten me as a foolish slut, my Lord,
> you come to threaten me
> Even as I go to pick the korai around the lake, my Lord,
> around the lake,
> you should know I am heavy with child, my Lord,
> you should know.[48]

In the Andhra and Karnataka regions, the basket weavers belonged to the Medara caste. There is a reference to women basket weavers whose work is known as *kantakara vritti* in the fifteenth century text *Kreedabhiramamu* of Vallabharaya.[49] The work was apparently very low paid and the Medara women were poor because *Kreedabhiramamu* says that many of these women basket weavers took to prostitution out of compulsion.[50] Besides making baskets and mats out of *korai*, baskets were also made of *panai olai* or palm leaf and bamboo called *moongil*. The old type of baskets and coconut leaf or palm leaf umbrellas are still a common sight in South India and these are still being made largely by women. Basketry is the specialised craft of the *Parava* (fisherwomen) of Kerala and in the Tamil country the women of Ramanathapuram are known today for their intricate basket work.

A craft in which women are said to be as skilled as men is pith work called *netti* in South India. The plant known as *shola* has soft flexible stems with spongy cellular tissues in them and also in the branches making it ideal for crafting things. With *netti* flowers, toys, hair decorations especially coronets, temple garlands etc., were made. The reference to pith work is found in the Sangam and post-Sangam literature. The *Silappadhikaram* refers to the *netti* work done in Pumpuhar and the sale of *netti* products in the markets of the bustling port town of Kaveripumpattinam. *Netti* working has by and large been a craft dominated by women. In Karnataka among the Gudigaras, the men carve wooden figures and idols while women engage in pithwork. The *Shimoga District Gazetteer*[51] which draws largely from the imperial gazetteers and manuals of the nineteenth century such as John Sturrock's *District Manual of South Kanara,* give very interesting details about women netti workers. It says that the women gathered the pith from around Shimoga and Shiralkoppa near the tank area and made beautifully crafted items with just a knife and a pair of scissors. It is said that a woman could make about 20 *venis* (pith strung together of an arm's length called *muzham*) and 20 garlands in a month. The work was done by women in the verandahs of their homes. The account is categorical in stating that pith work was very poorly paid in comparison to sandalwood carving or ivory carving which fetched big sums of money in the market and were entirely male preserves.

Conclusion

Traditional crafts in pre-colonial India was far from being a monolith in terms of its structures, developments or responses to colonialism. In fact different crafts were poised differently in terms of their technological development. Nor would it be appropriate to regard the colonial intervention as the sole marker for understanding the nature of traditional crafts. This is not to deny the impact of colonialism in fostering discriminatory development of the Indian economy but to draw attention to the richly textured differences within the traditional crafts ranging from pottery to weaving to the intricacies of icon making and metal working.

What does emerge powerfully from a study of traditional crafts and technology in pre-colonial peninsular India, is the symbiotic link that exists between crafted objects and craftpersons. In striking contrast to the standardisation associated with the industrial mode of production, traditional crafts were characterised by asymmetry. In the colonial period, a recurrent phrase in the English Company records, is the injunction to the Company Agents to "Please make them weave to 'the perfection of the pattern'".[52] 'Perfection of the pattern' went against the very ethos of the craftsmen who could not kill their individuality and creativity through monotonous repetition. Unfortunately if a weaver or craftsman digressed from the muster provided by the Company Agent, what he produced would be summarily rejected. As he worked on the basis of an advance, this could push him to desperation, even to the extent of committing suicide.

Asymmetry was and is an important aspect of the traditional craftsman's persona. Symmetry on the other hand is in a sense artificial, as it breaks the natural beauty of asymmetry. I have an example of a saree produced by the Saurashtra weaving community from Uraiyur, an ancient weaving centre going back to the Chola period. The craftsman had put together two absolutely asymmetrical objects – one an East India Company ship and the other, placed by its side, a traditional flower comprising the conventional pattern of a South Indian saree. What connected these diametrically opposed motifs was only the connection in the craftsman's mind between colonialism and his own traditional weaving craft.

Kamaladevi Chattopadhyaya, who was instrumental in salvaging Indian crafts from the twin onslaughts of colonialism and consumerism, beautifully expresses the link between technology, society and traditional crafts:

> craftsmanship grew from the village community, its joys and burdens, the change of seasons, the memories filled with song and verse, legends, myths and local romances, from the core and substance of their daily existence … .[53]

Notes

1. Ca utta University Lecture Series, reprint 1938.
2. Barun De, *Problems of the Study of Indian History: With Particulars Reference to Interpretations of the Eighteenth Century*. Calcutta: Centre for Studies in Social Sciences, 1988.
3. Frank Perlin in his essay "Proto-Industrialization and Pre-Colonial South Asia", *Past and Present*, No. 98, 1983.
4. Richard B. Barnett (ed.), *Rethinking Early Modern India*. New Delhi: Manohar, 2002. A thought provoking review of this book is by Amar Farooqi which has been published in *The Book Review*, Vol. XXVII, No. 3, March 2003.
5. Seema Alavi (ed.), *The Eighteenth Century in India*. Delhi: OUP, 2002.
6. Vijaya Ramaswamy, "Interactions and Encounters: Indian Looms and Craft Traditions Abroad – A South Indian Perspective", in A. Rahman (ed.), *Project of History of Science, Philosophy and Culture: India's Interaction with China, Central and West Asia*. Delhi: OUP, 2002, p. 438 and pp. 428–44.
7. To give a couple of examples of studies of this nature, see Eric Stokes, "The First Century of British Colonial Rule in India: Social Revolution or Social Stagnation?" *Past and Present*, Nos. 58–61, 1973, pp. 136–60 for a general survey. Sushil Chaudhury's recent book, *From Prosperity to Decline: Eighteenth Century Bengal*. Delhi: Oxford University Press, 1995, updates his many years of research.
8. V.I. Pavlov, *Historical Premises for India's Transition to Capitalism*. Moscow: Nauka Publishing House, 1979, Section Two, pp. 163–214.
9. See Vijaya Ramaswamy, "South Indian Textiles: A Case for Proto-Industrialization?" in Deepak Kumar (ed.), *Science and Empire*.

New Delhi: Anamika Prakashan, published by the National Institute of Science, Technology and Development Studies, 1991, pp. 41–58.

10. This line refers to the widespread notion among the weaving communities that the weavers obtained their first thread from the lotus like navel of Vishnu (the umbilical cord!).
11. The poem is ambiguous but may even indicate the early method of pattern weaving by manual lifting of the threads.
12. The translation of this poem which can be better described as transcreation, is mine. See my article "Weaver Folk Traditions as a Source of History", *Indian Economic and Social History Review*, Vol. XIX, No. 1, January-March 1982, pp. 55–56 for the Tamil text of the poem.
13. South Indian Inscriptions (henceforth SII), Vol. XVII, No. 603, Nagapattinam Talaq, Tanjavur district.
14. Annual Report of Epigraphy, 479 of 1908 from Uyyakondan Tirumalai, Tiruchchirapalli Talaq and district.
15. Annual Report of Epigraphy (henceforth A.R.E.), 575 of 1909. The inscription which is on a Naga pillar in the courtyard of the Chennakesavaswami temple records gift of land to the temple of Adityesvara.
16. South Indian Inscriptions (henceforth S.I.I.), Vol. VI, No. 439.
17. S.I.I., Vol. III, Part 3, No. 205.
18. A.R.E., 331 and 332 of 1936–37. The first is from Repalli in Tumkur district of Mysore dated AD 882 and the second record from Palnad taluq of Guntur district is dated in the 7th century AD.
19. A sixth century inscription from Badami written in archaic Telugu refers to Vidyadharan Kalgabharana Achrya S.I.I., Vol. XII, No. 23 A.
20. S.I.I., Vol. XVII, No. 603, Tiruvarur, Nagapattinam Talaq, Tanjavur district.
21. A.R.E., 479 of 1908 from Uyyakondan Tirumalai, Tiruchchirapalli Talaq and district.
22. A. Appadorai, *Economic Conditions in Southern India, 1000–1500 AD*. Vol. I, Chennai: University of Madras, 1936, pp. 357 ff.
23. Karl Marx in *Capital*. Vol. I, the English translation of the third German edition by Moore and Aveling, (eds.), by F. Engels, (Moscow, 1958), pp. 339–40.
24. Ibid., p. 340. This point has been discussed in my article, "The Masterweavers in South Indian Textile Production", *Journal of the Economic and Social History of the Orient*, Vol. XXVIII, Part III, Oct. 1985, p. 297.
25. This article was first published in the *Socialist Digest*, No. 6, in 1972. See specially p. 105 ff.
26. K.N. Chaudhuri, *The Trading World of Asia and the English East India Company*. Cambridge: Cambridge University Press, 1978, p. 274.

27. K.N. Chaudhuri, "The Structure of Indian Textile Industry in the Seventeenth and Eighteenth Centuries", *Indian Economic and Social History Review*, 1974, p. 152.
28. K.N. Chaudhuri, ibid., pp. 180–81.
29. The English Factories in India (henceforth EFI) William Foster (ed.), 1606–69, 13 Volumes, (Oxford 1906–27), letter dated 25th June, 1929, p. 342.
30. Ibid., 26th Oct., 1646, p. 52.
31. Ibid., 4th April, 1954, p. 256.
32. This point was discussed at some length in my book, *Textiles and Weavers in Medieval South India*. Delhi: OUP, 1985, pp. 151–52.
33. *Parasara Madhaviyam*. vide A. Appadorai, *Economic Conditions in Southern India*. Vol. I, Chennai: University of Madras, 1990, p. 236.
34. *Parasara Madhaviyam*. Vol. I, p. 432 cited in ibid., Vol. I, p. 235, fn. 616.
35. Hamilton Buchanan, *A Journey from Malabar through the Countries of Mysore, Canara and Malabar*. Three Volumes, T. Cadell (ed.), Vol. I, London: W. Davis and Black, 1807, p. 100.
36. Colonial Alexander Walker's comment on the Malabar plough is from a detailed report on the subject, the full text of which is published in Dharampal's book, *Indian Science and Technology in the Eighteenth Century*. The above passage is taken from Claude Alvares' book *Homo Faber: Technology and Culture in India, China and the West – 1500 to the Present Day*. New Delhi: Allied Publishers Private Ltd., with the Centre for Developing Societies, 1979, p. 50. This book also contains an excellent discussion on the issue of appropriate technology.
37. Ibid., p. 49.
38. Vijaya Ramaswamy, "Textiles and Weavers", op. cit., chapter V. The revised edition of this book is now under publication.
39. Elijah Hoole, *Madras, Mysore and South India*. London: 1844.
40. *Visvakarma Puranam*. Mackenzie Manuscripts, Wilson Collection, No. 72, pp. 6–7.
41. Although multiple oral versions of this community tradition exist, an excellent textual analysis is to be found in Jan Brower, "The Story of the Magnetic Fort", published in his book *Makers of the World*. Delhi: OUP, 1995. Also see "The Story of the Magnetic Fort", by the same author in A. de Ridder and J.A.J. Karremans (eds.), *Tradition in Structural Anthropology*. Netherlands: E.J. Brill, 1987. This is not to say that this is the only authentic version since other oral traditions differ from it in significant respects.
42. Hamilton Buchanan, op. cit., Vol. I, Chapter VII, p. 42.
43. J. Brouwer, "A Matter of Liminalities: A Study of Women and Crafts in South India", *Man in India* 67, No. I, March 1987, p. 6.

44. The details of women's participation in the handloom sector during the Sangam age (roughly from circa third century AD to third century BC, is to be found in my article "Aspects of Women and Work in Early South India", *Indian Economic and Social History Review (IESHR)*, Vol. XXVI, No. I, January–March, 1989, pp. 85–87.
45. Sharanacharitramu, p. 223 cited in J.K. Kamat, *Social Life in Medieval Karnataka*. New Delhi: 1980, p. 119.
46. For the powerful poetry within the Virasaivite tradition by women spinners, see my book *Divinity and Deviance: Women in Virasaivism*. Delhi: OUP, 1996, pp. 53–54.
47. Vijaya Ramaswamy, "Women and Work", *IESHR*, op. cit., p. 86.
48. Ibid., p. 88. The translation from the original Tamil is mine.
49. The published version of Kreedabhiramamu in Telugu was edited by Vetturi Prabhakara Sastry – Muktiyala, 1952. See Verse 67. A recent translation of the same text has been brought out by the Oxford University Press.
50. Ibid., Verses 101 to 105.
51. Published by the Government of Karnataka, Bangalore, 1975, pp. 207–09.
52. This phrase occurs often in the Company correspondence but one significant instance is the letter dated 1696 regarding the musters to be sent to the Coromandel weavers through their Agent Mr Ongley – India Office Archives, Letter Book IX, 466. This is cited in John Irwin, "Indian Textile Trade in the Seventeenth Century: South India", published in *Studies in Indo-European Textile History*. John Irwin and P.R. Schwartz (eds.), *Calico Museum of Textiles*. Ahmedabad, India: Calico Museum of Textiles, 1966, p. 37.
53. Kamaladevi Chattopadhyaya papers, Nehru Memorial Museum and Library cited in Reena Nanda, *Kamaladevi Chattopadhyaya: A Biography*. New York: OUP, 2002, p. 126.

9 Glimpses of Craft Technology in Kashmir

Rattan Lal Hangloo

> There is scarcely a thing which he (Kashmiri craftsman) cannot do. As a fine craftsman, he may have a few equals in the world but probably none superior to him.
>
> — Walter Roper Lawrence[1]

From time immemorial Kashmiris indulged in craft production to supplement their sustenance derived from one crop-based agricultural economy.[2] The excavation at Burzahom near Srinagar in Kashmir revealed that people used saw edged bone objects, scrapes of horn with broad butt and polished sharpened working edges and stone pounders with grinding marks and smoothed surfaces and stone querns in antiquity (picture).[3]

Between AD 78 to 3rd century AD the handicraft production of Kashmir is listed as the major chunk among various items of Indo-Central Asian trade.[4] There are references to swords and razors during King Mihirakula's time (AD 51).[5]

One of the earliest documentary evidences about craft items and the tools involved in their production is provided by *Nilmata Purana* of seventh century AD.[6] The author of the *Nilmata Purana* records that while adoring God the expressions of garments like glowing gold and several other items were produced by Kashmiri craftsmen.[7] It refers to various tools used by them. Articles of dress allude to arts of spinning, weaving, dyeing and washing. There is also mention of ornaments, vessels made of gold and silver, silver tools and the plough pointing to the existence of crafts men, gold and silver smiths.[8] There are frequent references to earthen and wooden pitchers, wooden seats, leather, shoes, clothes, umbrellas, sweat meat, cakes, weapons, bells, toys, pottery, carpentry, painting, perfume, wine, ornaments, leatherwork, musical instruments particularly *Tantri* instrument[9] White clothes and bed sheets are also referred to in *Nilmata Purana*.[10]

After bathing the *linga,* i.e., *(Shaiva image)* a woollen covering had to be placed over it.[11] On *shraddha* ceremony a new silken cloth was to be given to the priest.[12] These references clearly indicate that the varied nature of artisanal production and technology deployed by the craftsmen was very well established by the time *Nilmata Purana* was written in seventh century AD. All this shows the existence of a vibrant artisanal community in Kashmir. It is indeed important that compared to other parts of India the variety of crafts is indeed phenomenal in number which is directly correlated to the artisans and references in literature and also the complex typology of tools. This shows the high degree of specialisation and nature of intricate and subtle operations.

Besides these pieces of evidence the various rituals mentioned in *Nilmata Purana* are also associated with the variety of crafts produced during the period. It is stated that in the month of *Ashada* (June/July) the statu of *Keshava* is required to be made out of clay, stone, gold, and silver, brass copper or wood.[13] There are references to rituals performed by artisans while worshipping their tools and *vishwakarma* (the originator of all crafts) on specific occasions[14] Artisans were directed to go on fast on a particular day to worship *Bhadrakali (mother Goddess)*.[15]

From ancient times artisans were held in great esteem in Kashmiri society. They played an important role during the coronation ceremonies of the kings. They decorated their shops and plazas on *Kartika Amavasya*. *Nilmata Purana* states, "*Vishwakarma* the originator of all crafts should also be worshipped oh, Brahaman each of these namely the weapon, the conveyance, the umbrella the seat, the emblem and the drum should be worshipped separately with scents, garlands, paints, offerings of lamps, incense and eatables".[16] This also corroborates the fact that Kashmir stood at the periphery of two great civilisational entities, the Indo-Aryan and the crossroads of Central Asia. This factor and consequent entry of loan words and indigenisation of fresh ideas saw that the crafts here never became ossified. A simple example of this is seen in the fact that newer ideas and techniques came but the artisanal communities were not replaced by migrants, only the products. This shows the adaptability of the craftsmen. As an extension it should also be pointed out that the patronage patterns were conducive to this spread, a fact not highlighted by Marxist historians despite their significant contributions to Indian history.

There is mention of an iron board studded with razors, swords, knives and sharp instruments.[17] He refers to swords, lamps, glowing fire pans.[18] In other parts of India craft production received a great boost with the growth of surplus but in Kashmir the regions, geography and ecology supported only the meagre agricultural economy therefore the craft produce constituted an important supplement from the very inception of material culture.[19]

The very fact that Pravarasena II (who ruled for four years and his date of accession is only given in *Laukika* 3186 by Kalhana) got the great bridge *Brhatsetu* built on the *Vitasta,* also referred as *nausetu,* speaks of the technology of making bridges in early Kashmir.[20] It was constructed by using boats as pillars to hold the weight and was known as boat Bridge.[21] These bridges continued up to early 20th century with repairs at different intervals.

"These bridges", wrote Baron Hugel in AD 1835, "were found already laid across the river by the Muhamadans which gives them an antiquity of at least 500 years. Since the dominion of last Hindu sovereign, or more correctly, of the last queen of Kashmir, Rani Kotadev is (popularly known as Kota Rani) which, according the *AyinAkbari* (*Ain-Akbari*), terminated in 1364, the last partial restoration was undertaken by the governor Ali Mardan Khan in the reign of Emperor Jahangir (Shah Jahan)".[22]

When M.A. Stein visited Kashmir in the 19th century he also noticed these bridges. He stated that the number of these bridges has remained unchanged for, at least, five hundred years.[23] The Hindus were most unlikely unskilled in building stone bridges because we have many temples constructed of stone which display the fineness of art and technology involved in their make. But why they constructed boat bridges can perhaps be attributed to abundance of wood which could be easily transported from forests through rivers.

These wooden bridges did not involve intensive human labour and could also be dismantled easily during battles ,which were a perpetual feature in Kashmir. This technology of wooden bridge building was very much in use in pre-Sultanate period before the 13th century but it became common from the Sultanate period (AD 1339–1540) and continued down to early 20th century.[24]

The fact that Kashmirees were familiar with use of stone in construction activity and the masionary styles becomes clear from evidence available in *Rajatarangini* and other ancient texts. While describing the constructions of Pravarasen II, Kalhana says, "There are mansions which reach to the clouds and ascending which one sees the earth, glistening in the reign at the close of the summer and covered with flowers in the month of *chaitra*".[25]

Kalhana refers to the construction of four lakh stone buildings in the town of Lohara.[26] Construction of *Amritbawana* and *Nandavana,* both with wonderful appearance is also recorded.[27] There are references to lofty enclosure *prasadantar* and many building and towns like *Lokpunya, Parihaspura, Cakrapura, Surapura, Avantipura. Shurapura, Pattana, Padmapura, Vijayeshvara* stone dams and embankments, dykes, *viharas, mathas, chaitiyas, setu's* (bridges) etc.[28]

The famous sculptor of Pravarasena II was Jaya.[29] One of the images of *Shiva* in the city in *Sharitaka* was known after him as Jayaswamin.[30] Ksemendra's *Narmamala* and *Rajatarangini* of Kalhana refer to learning, lofty houses, prosperous markets in ancient Kashmir.[31]

Once Pratapditya (the date of his accession according to *laukika* is 2896) was invited to his house by a merchant he found lamps formed of Jewels (*manidipika*).[32] There are references to leather workers (tanners).[33] Kamalavati the queen of the great king Lalitaditya Mukhtapida (AD 649–736) also established Kamlahatta (market of Kamla). The premier markets were *puranadishthana, Parihaspura, Hushkapura Jushkapura* and *Pravarpura*[34] During his period Lalitaditya put into the image of *Mukhteswara* eighty four thousand *Tolikas* of gold and collected as many thousand *palas* of silver that pure minded king made the famous statute of *parihaskeshava*.[35] He is said to have planted a series of machines at Chakradhara to draw water from *Vitasta*.[36] The bronze and copper craftsmanship of Kashmir attained its artistic peak under Laltadityas rule (AD 649–736).

King Lalitaditya (AD 649–736) got images of gods and goddesses made of gold and silver for various shrines. His queen Kamalavati got large silver images of *Kamlakeshava*.[37] King Jayapira wore Keyuras (an ornament worn on the upper arm).[38] Avantivarman (AD 855–883) also patronised the arts and crafts.[39] During Samkarvarman's time (AD 883–902) there is a clear reference to preparation of woolen clothes at pattan in his time.[40] The area of *Pattana* was very famous during the reign of Samkarvarman (AD 883–902) for weaving of woollen clothes.[41] There is reference to use of coarse woollen *(phiran)* cloaks used by people in Kashmir even as far back as AD 939.[42] It establishes the fact that there was increasing material activity because of the craft production which not only increased trade but also gave a boost to craft technology from time to time. The existence of markets is well-known and also the existence of exchange of goods. But many of the writers come close to saying that there were 'trade networks', but do not use this category. This in fact is a historiographical and not a historical problem, a fact that is revealed by the methodological bias of the earlier historians.

King Harsha (AD 1054–1101) is said to have given due allowance to garment-making, ornaments and gem craft and jewellery.[43] King Harsha introduced various fashions. During his tenure as prime minister he started wearing a short coat and this would not have been possible without the availability of required knowledge and technology.[44] There was an increasing interaction between Kashmir and Central Asia.

Repeated references occur in texts about *Dinara, Hundika, Dinarojjama* (bond of debt for cash) *Dhanyhojjamacrica* (bond of debt for Danya) show that Kashmir had well developed science of monetary system.[45] There are references to nipping of money lenders by Ksemendra[46] (commutation of kind into money) or minting of coins would not have been possible without technology. The items of craft that were available included potters wheel, water wheel, wood work, copper and brass work.[47] The later period of *Rajatarangini* also shows the existence of silver articles like golden and silver figurines, golden couch, parasol of gold, golden statutes, golden sticks and golden armour of Keshva.[48]

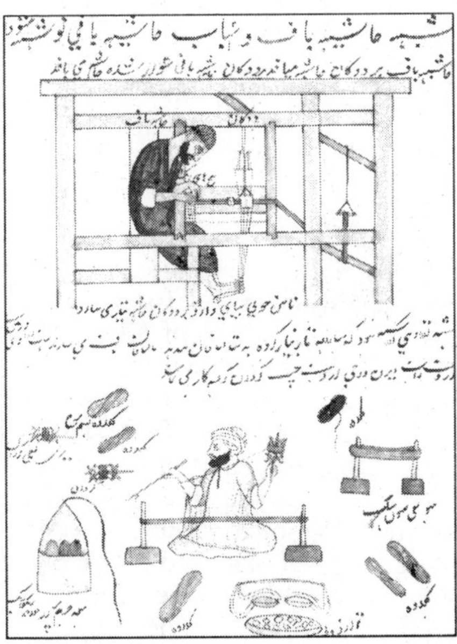

Figure 1: The shawl weaver and his assistant. Also seen are the varieties of wool and tools

Kalhana says, "On sandy bank of Vitasta, he (King Nara) built a town where markets were kept full of supplies by the high roads (leading to it) where coming and going of ships (boats) gave splendor to the river".[49] When the king Kalasa (AD 1063–89) wished to put a gilt parasol on the (temple of *Shiva Kalasa*) artist was brought from the *Turusuka* country.[50] It was under King Kalasa (AD 1063–1089) that a Turkish mechanic was invited to help in plating copper with gold.[51]

During the time of Jaisimha (AD 1128–49) market buildings were given as gifts by Kings to persons of their choice.[52] Even the temple *purohitas* were granted this favour.[53] Grant of shops to individuals and shrines is mentioned in the inscriptions.[54]

From the early times many towns and buildings came up in Kashmir. Wooden structures were very prominent. Masonry work was rare but it did constitute an important item of building technology in 15th century. Budshah was a great builder and town planner; his reign was also characterised by continuous building activity. *Jonaraja* tells us about the appointment of Ruppya Bhandapati (a proficient architect) as the superintendent of buildings who supervised among others the construction of *Zainlankh* (an Island) with a complex buildings and the founding of towns and cities such as *zainkota, zainapattan, surtanpura and zainkundala, zainkadala*.[55]

Figure 2: Shawl menders and their tools

During his period Damara Kancha who was superintendent of the city also built a stone bridge one *Krosha* (roughly a mile in length) in length on the road which was difficult to traverse.[56] Rupyabhanda built *suratranapura (Sultanpura)* in Karamraja (*kamraz*) and also repaired dilapidated buildings.[57]

There are references to specimens of wooden architecture in Zain-ul-Abidin's time. (AD 1420–1470) The most magnificent wooden structure was the one constructed in Naushar. (in Srinagar) consisting of twelve storeys of fifty rooms each, with halls and corridors. In its massiveness it was one of its kind in the eastern world.[58] The other buildings reflecting wooden architecture were in the town of Suyapura which were gutted in a fire.[59] A number of epitaphs on the graves situated in the compound of Sultan Zain-ul-Abidin (AD 1420–1470) belong to men of different occupations among whom one finds masons one of whom is referred to as Hussain Sangtarash.[60]

The other important specimen of this wooden structure was *Jamia-Masjid* built originally in AD 1400. This building was also gutted by fire three times but was rebuilt each time which led to certain changes in its style of construction. Its original plan was designed by Sadr-ud-din who had come from Khurasan.[61] The mosque had 370 columns when it was built originally in Sultan Sikander's time.[62] The whole structure is enclosed by a lofty outer wall of masonry with projecting entrances on three sides. Its striking architectural features are the elegantly recessed portico, arched arcades, lofty archways, and the pyramidal roof surmounted by a spire. But the singularly imposing aspect is a series of lofty columns of deodar that surround the entire structure. The columns are 378 in number ranging between 7.62 and 15.25 metres in height.[63]

Only a few buildings in masonry style have survived from Zain-ul-Abidin's time. They are the tombs of Zain-ul-Abidin's mother and the tomb of Mir Sayyid Mohammad Madni These structures show the affinities to contemporary brickwork and glazed tile architecture of Persia.[64] Also these structure clearly reflect architectural conversion because the plinth and enclosure wall of the building are that of an ancient Hindu temple.

This certainly indicates that the art of architectural conversion did exist even before the Sultanate period.[65] The ancient appearance of stones used and the Hindu features such as mouldings of plinth and the section of coping, unmistakably prove it to be so.[66] The superstructure made of brick in accordance with Islamic needs is the work of the Sultan. Standing between fourth and fifth bridges in Srinagar, the tomb of Zain-ul-Abidin's mother is square in shape.

The four corners are straightened to constitute sides for rectangular projections. Based on a singular pattern, the super structure consists of a single centrally situated chamber with projections recessed internally at the angles.[67] It is surmounted by four cupolas of equal size resting on rectangular projections and a large towering dome attached to these.

It is noteworthy that the original plan of the Hindu structure facilitated its transformation into a "conventional Muslim tomb" without much difficulty. Each wall of this structure has an archway set into it. There is an iron chain inside the building, hanging from an iron plate, suggestive of a Hindu influence. The inner doorway is so designed that its upper portion gives the impression of a horseshoe arch. The upper important features of the tomb are" the glazed and moulded blue bricks, which are studded at intervals in the exterior walls, the semi circular brick projections on the drum of the main dome, and the mould, string-courses and sunk panels on the drums of cupolas.[68]

The design and execution of this tomb implies Persian technological influence. Built on the stone plinth of an ancient temple, the mosque has a brick superstructure consisting of four walls externally decorated with trefoiled brick niches. It would be pertinent to point out here that the trabeate system of architecture that was current in the Buddhist-Hindu period continued even later and there was no wholesale replacement pointing out to the fact that there was co-existence of art and architectural traditions and this was seen not in terms of religious categories but as pragmatic structures. This is mirrored in literature in the use of synonyms for different contexts, e.g., knife, axe and many other items. In Kashmir the knife is used for curing the skin diseases and warding of evil spirits but at the same time for many other kinds of work. Same is the role of axe.

In the construction of the porch some carved Hindu columns have been used. The mosque is a self contained structure with a hall measuring 9.68 metres square internally. one of the constructional features are the pilasters of walls and niches, spandrels of arches, cornice, capitals, eaves, dentils and metopes, beautiful tracery work and open work carving. The entablature and capitals are purely of Hindu style.[69] The structure has a pyramidical roof covered with earth which as the remnants show, was once surmounted by a spire.

Two features of the mosque are particularly noteworthy. First, the wooden doorway is elaborately carved and constitutes a fine specimen of the art of wood carving in Kashmir. Secondly the ceiling of chamber is finely decorated with thin pieces of wood. Worked into geometrical patterns and beautifully painted with flowers in

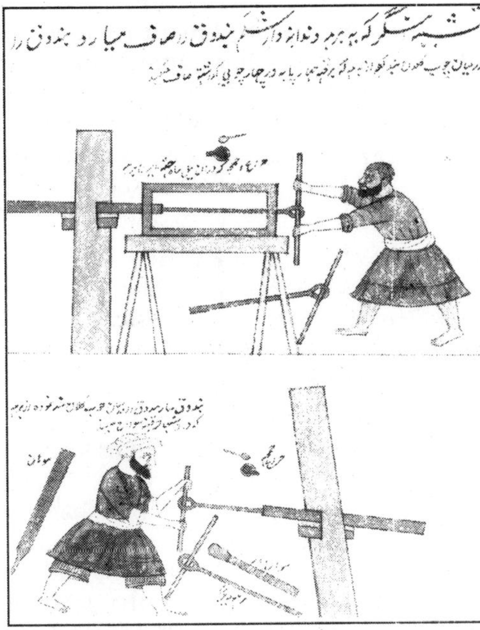

Figure 3: Ironsmiths at their workshop while making guns

different colours. It is called *Khataband*[70] standing on four multi-faceted wooden columns. The ceiling is designed into nine squares of equal size with the middle one bulging upwards into a dome.

The whole structure is crumbling. The tomb of Madni is also built on the plinth of Hindu temple. The superstructure is made of brick generally measuring 21×15×3cm. In the inner chamber of the tomb two Hindu columns stand. The chief attraction of the building is its glazed tile work and it clearly has Persian technological influence. The tiles used are square in shape, each side measuring 19.5cms. Rectangular tiles are used on the borders only. The colours are yellow, indigo brown, dark.[71]

In the Sultanate period wood formed the bulk of architectural material. The Sultan encouraged artists proficient in different skilled.[72] During Shahmir Sultanate (AD 1339–1540) the existing sciences and technology made a great progress in Kashmir because of the patronage extended by the Sultans. This process was further strengthened by the migration of Sayyids and a variety of people accompanying them from various regions of Central Asia. Zain-ul-Abidin (AD 1421–72) brought craftsmen from Iran and other parts of central Asia. He gave particular impetus to manufacturing of hand made paper, shawls, embroidered tapestry, silk weaving and wood carving.[73] He is said to have provided scholarships to those who went to Central Asia and some other countries to learn new techniques and designs in different arts.

The ruler of Gujarat, Sultan Mahmud also sent to Sultan Zain-ul-Abidin (AD 1421–72) a celebrated varieties of fabrics as gift to Kashmir. The varieties mentioned are towel cloth (*catefa*), camlet (*sof*) and some silken stuff brocaded with gold (*saqlat*).[74]

Dugarseha, the ruler of Gujarat also sent to Sultan Zain-ul-Abidin popularly known as Budshah (AD 1421–72) a couple of valuable treatises on the sciences of music.[75] One of these works entitled *Sangitchudamani* contained rules related to fine arts, including drama and music.[76]

Budshah also received some technical assistance.[77] Budshah had very close cultural contacts with Afghanistan, Persia, various parts of Central Asia, Gilan, Egypt and Mecca which would have also included technological exchanges.[78] Budshah is said to have sent cups of glass or crystal among other wonderful products like saffron, paper, musk and shawls to Mirza Abusaid (AD 1452–1467) of Persia which certainly indicates that the technology of glass, paper and shawls did exist in Kashmir.[79] The carpet weavers from Samarkand introduced carpets for the first time to Kashmiris.[80]

Mirza Haidar Dughlat says, "In Kashmir one meets with all those arts and crafts which are in most cities uncommon, such as stone polishing, stone cutting, bottle making, window cutting (*tabdan tarashi*), gold beating etc. In the whole *Mavar-u-Nahar* (Transoxiana) except in Samarqand and Bukhara these are nowhere to be met with, while in Kashmir they are abundant."[81] Mirza Haidar attributes all these developments to Sultan Zain-ul-Abidin. True that Zain-ul-Abidin revived and patronised many of these crafts but many of these existed in Kashmir prior to his regime as well.

Zain-ul-Abidin is credited with the introduction and revival of several new industrial crafts such as spinning, weaving, carpet making, book binding, papermaking, stone and glass works, plume making, boat making, window cutting, gold beating and several others. We also have evidence of professions of butchers, scribes, calligraphists, firework makers, miners, brick layers, potters, goldsmiths, physicians, gold washers, burja leaf makers, black smiths, brewers and wine vendors, shawl weavers masons, carpenters.[82] The preference for the wooden craft items of Turkish carpentry was so high that even the best carpenters are associated with Turkish carpentry (*Torka Chan*) even to this day.[83] All these details testify to the fact that craft making technology was an integral part of Kashmir's material culture from the very antiquity.

Figure 4: Sawholders and their tools at their workplace

Notes

1. W.R. Lawrence, *Valley of Kashmir*. London: Henry Frowde, 1895.
2. Because of the nature of Kashmir's geography and ecosystem, the economy of the region was never predominantly given over to agriculture, therefore, the tendency of supplementing the one crop economy by craft production and pastoralism remained a permanent characteristic of the economy of Kashmir from the very inception of material culture in the Valley. For details see M.A. Stein, *The Ancient Geography of Kashmir*. Calcutta: *Journal of Asiatic Society of Bengal*, 1899 (reprint), 1977, pp. 117–25.
3. B.B. Lal (ed.), *Indian Archaeology 1968–69: A Review*, published by Archaeological Survey of India, Government of India, New Delhi, 1971, p. 10. See also Plate XII B of the same report.

4. V.M. Masson and V.I. Sarianidi, *Central Asia: Turkmenistan before the Achaemenids*. London: Thames and Hudson, 1972, pp. 124–28. See also R. Ghirshman, *Iran: From the Earliest Times to Islamic Conquest*. Great Britain: Pelican Books, 1951, pp. 62–72.
5. Kalhana; *Rajatarangini* (Sanskrit) English translation by M.A. Stein 2 Vols., London, 1900 (reprint), 1961 and 1979, Vol. I, Book I, pp. 43–47, footnote 289 (henceforth Kalhana *R.T.*).
6. *Nilmata Purana* (Sanskrit) English translation by Ved Kumari Gai Jammu and Kashmir Academy of Art, Culture and Languages Srinagar 1968 (reprint), 1988, 2, Vol. I, pp. 107–08.
7. *Nilmata Purana*. Vol. II, v. 62, pp. 17–18.
8. Ibid.
9. Ibid.
10. Ibid., Vol. I, v. 172262, 323, 373, 425, 427, 444–45, 448, 452–54, 458, 475, 483–84, 509, 541, 558, 567–68, 671–72, 685–86, 690, 723, 756, 759, 814. Vol. II, 64, 492, 578, 819–20, p. 45, 110, 114, 125, 138, 142, 144–45, 170, 174–75, 184, 192–62.
11. Ibid.
12. Ibid.
13. Ibid.
14. Ibid., Vol. II, p. 164, 208, vv. 646–47, 819b–820a.
15. Ibid.
16. Ibid., Vol. ii, vv. 646–47, p. 164.
17. Ibid., Book I, vv. 312–16, p. 47.
18. Ibid., Book III, vv. 50, 171, pp. 76–86.
19. Stein, *Ancient Geography*. op. cit.
20. Kalhana; *R.T.* Vol. I, p. 103, 136.
21. Kalhana; *R.T.* Vol. I, p. 103. See also reference of begs material. King Harsha (AD) also constructed great bridge *mahasetu* with boats. Ibid., Book V, ii, p. 388, v. 1549.
22. C.B. Hugel, *Travels in Kashmir and Punjab*. London: John Petheram, 1845, p. 117. See also Sufi, *Kashir*. Vol. II, p. 521.
23. Kalhana; *R.T.* Vol. II, p. 449.
24. For details see R.K. Parmu, *A History of Muslim Rule in Kashmir*. Delhi: People's Publishing House, 1969.
25. Ibid., Book III, 359, p. 104, 227.
26. Ibid., Book I, 86, p. 15.
27. Ibid., Book III, 359, p. 104.
28. Ibid., Book iv, v. 188, 190–213, 14, 512, pp. 141–42. The town of Suyapura was founded by Avantivarman (AD 855–883). Ibid., Book v. p. 190–91, v. 39–44, p. 198, v. 91, p. 199, v. 101, p. 201, v. 114–15, p. 213, v. 213, Book vii, v 409, p. 301, Book VII, v. 938, p. 341.

29. Ibid.
30. Jonaraja, Vol. I, p. 53.
31. Kalhana; *R.T.* Vol. I, p. 10. Book I, v. 42, p. 123.
32. Ibid., Book iv, vs. 11–15, p. 121.
33. Ibid., Book Fourth, vv. 55–62, p. 125.
34. Ibid., Book iv, 208, p. 143. See also P.N.K. Bamzai, *A History of Kashmir: Political, Social and Cultural from Earliest Times to Present Day*. Delhi: Metropolitan, 1962, p. 229.
35. Ibid., Book Fourth, vv. 200, 201, p. 142, with many a thousands of *prathas* of copper he made the glorious statute of the great Buddha (*Brahadbuddha*) which reached up to the sky. Ibid., Book V. 203, p. 142.
36. Ibid., Book IV, v. 188–98, p. 140, 141. See also Jonaraja, op. cit.
37. Ibid., Vol. I, Book IV, v. 208, p. 143.
38. Ibid.
39. Ibid., Vol. II, Book V, v. 162, p. 207.
40. Ibid., Vol. I, Book v, v. 162, p. 207.
41. Ibid.
42. Ibid., Vol. I, Book V, p. 233, v. 461. See also Book VII, p. 395, v. 1645. See also Vol. II, Book viii, v. 857, p. 68.
43. Ibid. See also D.N. Saraf, op. cit., p. 40.
44. Ibid., vv. 921–24, p. 339.
45. *Lokaprakasha (Sanskrit) Kashmir Series of Texts* II, XVIII, p. 339. Srinagar 1898, see also *R.T.* Vol. II, p. 328, Note.
46. Kashmendra, *Kalavelasa* (Sanskrit) Kashmir series of texts research department library, Kashmir University, Srinagar, Vol. I, pp. 19–26.
47. Kalhana *R.T.* Vol. II, Book iv, v. 191, pp. 140–41.
48. Kalhana *R.T.* Vol. I, Book iv, v. pp. 205–08.
49. Ibid., Book iv, vs. 201–02, p. 142.
50. Ibid., p. 207.
51. Ibid., Vol. I, Book vii, v. 528, p. 311.
52. Ibid., Book viii, v. 3319, p. 259.
53. Ibid., p. 259.
54. Ibid., Vol. I, I, footnote to v. 3319, p. 259.
55. Jonaraja, Vol. III, pp. 94–95 and p. 127.
56. Ibid., Vol. III, p. 89.
57. Ibid., pp. 94–95.
58. *Tarikh-i-Rashidi*. pp. 429–30.
59. Srivara, stt. 34, 35, 41.
60. B.B. Lal (ed.), *Indian Archaeology1967–68: A Review*, published by Archaeological Survey of India Government of India, New Delhi, 1968, pp. 58–59.

61. Haidar Malik, op. cit., p. 44.
62. *Baharistan*. op. cit., p. 53. Percy Brown, *Indian Architecture Islamic*. p. 87.
63. Haidar Malik Chadura, op. cit., p. 44.
64. Percy Brown, *Indian Architecture*. Calcutta: University of Calcutta, 1942, p. 87.
65. Alexander Cunningham, *JASB*. Vol. XVII, Part II, 1848, pp. 245–46.
66. *Archaeological Survey Report*. 1906–07, p. 164.
67. R.C. Kak, *Ancient Monuments of Kashmir*. London: Royal Asiatic Society, 1933 reprint, p. 81 (henceforth R.C. Kak).
68. *Archaeological Survey Reports*. 1906–07, p. 164.
69. R.C. Kak, op. cit., p. 92.
70. Ibid.
71. Percy Brown, op. cit., p. 88.
72. Ibid.
73. D.N. Saraf, op. cit., p. 39.
74. Ibid.
75. Srivara *Rajatarangani*.1, vi, st. 14, p. 150, Nizamuddin Ahmad in *Tabkat-i-Akbari* (henceforth Nizam-ud-din, *Tabqat)* mentions him as Dungarsen, p. 659.
76. Nizam-ud-din, *Tabqat*. pp. 659–60.
77. Srivara, 1, vi, st. 25.
78. *Baharistan*. p. 45. Srivara 1, st. 26, *JASB*, Vol. XLviii, 1879, p. 284. Nizamuddin Ahmad, op. cit., p. 659.
79. Nizamuddin *Tabqat*. n. P. 659. Srivara, st. 22–24. Ain Vol. Ii, p. 383.
80. Sayyid Ali, op. cit., folio.15a.
81. Mirza Haidar Dughlat, *Tarikh-i-Rashidi*. p. 434.
82. N.K. Zutshi, Zain-ul-Abidin, pp. 115–51.
83. *Baharistan* p. 49. The preference for the wooden craft items of Turkish carpentry was so high that even the best carpenters are associated with Turkish carpentry *(Torka Chan)* even to this day.

10 Textile Technology in Medieval Orissa

A Case Study of Production and Export During Seventeenth and Eighteenth Centuries

K.N. Sethi

Production and distribution of economic goods in India in general and Orissa in particular was based on the co-existence and, at times, interpenetration of subsistence and a commercialised sector. As the bulk of the population lived in the villages and their needs for goods and services were satisfied through production for use and a network of reciprocal obligations, exchange accounted for a relatively small proportion of economic activity. Yet exchange of goods, found at virtually every level and sphere of economic life, was impressive in its magnitude and complexity. The dominance of subsistence-oriented production was modified by surplus and deficit necessitating multi-tiered and multi-faceted commercial activity.

Since capitalism fundamentally alters the relationship between agriculture and industrial production, with the increasing phenomena of non-agricultural or industrial production it is natural to look for the signs of the beginning of capitalism in conditions that obtained in the non-agricultural sector of the country. Although agriculture formed the main feature of the economic life of the people, there existed several crafts and industries in Orissa during the period under survey. It seems a considerable portion of Orissa's population earned their livelihood working as manufacturers of various kinds of non-agricultural products. The abundance of the products was such that even after the needs of the people were met fully, there was considerable surplus left for export.

I

Textile production was the most important sector of the Indians economy, after agriculture, in pre-colonial times. In the seventeenth century it also accounted for a significant share in the entire international trade. It has often been assumed that, crafts have no history in India, i.e., all the elementary tools, devices and practices were already in existence at the earliest imaginable times – "earlier even than the immigration of the Aryans"[1] – and that all developments since then have been of a minor or very secondary nature.

Textiles were most important, both in terms of value and volume, in the overall picture of Orissa's production during the period under study. The European Companies began to display interest in the Indian textile trade in the early seventeenth century for the purpose of bartering cotton piece goods in the Indian archipelago for pepper and spices. The direct trade in textiles between Europe and India developed as an essential by product of this 'earlier and more urgent necessity'. The most striking feature of the English East India Company's textile trade from Bengal and Orissa was a boom in export, which began, in the early 1680s under the stimulus of a rapid expansion in demand for *calicos* in European markets, and it continued, with the exception of a brief interruption due to the political turmoil.

The multiplicity of the types of textiles exported from Orissa makes its identification and proper division into different categories. It is not easy to identify some of them such as *umbers, mohmudiaties, atchabannies, abrowahs, bulcholls, coopes, doodamies,* etc. Notwithstanding this limitation, the piece goods exported by the Company may be divided into three main types: firstly silk piece goods, secondly mixed piece goods, that is, piece goods of mixed silk and cotton and thirdly, cotton piece goods, plain or painted. In addition, there was another category of miscellaneous goods consisting of quilts, tablecloths, plushes, velvets etc. Sometimes, however, some of the piece goods such as *rumals* and *lungis* could be of pure silk or cotton, or even of mixed cotton and silk.

Silk Piece Goods

The English knew silk piece goods by the term *taffatie* or *taffeta* and the Dutch termed it *armosijnen*.[2] The word vaguely implies fine cloth, usually of a silky and glossy quality. When the Europeans introduced the term into India, it became mixed up with Persian *tafta*, 'a glossy twist', already in use as a term for silk.

Orissa produced varieties of silk goods during the period under review. *Tassar* silk was available during the time of Master in

Balasore.[3] In AD 1680, the importance of Balasore as a silk manufacturing centre increased, when the English demanded silk and other goods'.[4] Master has mentioned about the offer to the merchants of Balasore to sell 12,000 pieces of silk *rumals* to the English East India Company.[5] The Company's requirement for 7,500 pieces of silk *rumals*, 6,000 pieces of *Herba Taffataes* and 20 bales of *tassar* or *herba* thread were to be provided at Balasore in AD 1681. In AD 1708, Alexander Hamilton referred to the production of silk cloths at Balasore.[6] In AD 1676, Walter Clavell wrote that Balasore, which was near the country of the Raja of Tillbrichrumbung, or Mayurbhanj, where the best quality and the largest quantity of Tester (*tassar*) or *herba* was procurable. Further, "*Ginghams, Herba Taffetas, Herba Lungis* (*Lungi* or loin cloth) and other sorts of *herba* goods" manufactured in neighbouring places could easily be brought to Balasore. Among other silk piece goods exported by the English Company were silk *rumals*, neck cloths and *atlases* woven mostly produced in Balasore area.[7]

Pipli was another centre where silk was manufactured during the time of Thevenot. William Bruton, in AD 1633, mentioned the production of silk textiles at Puri.[8] Jaleswar and Hariharpur were noted for their silk textiles during the time of Manrique.[9]

Piece Goods of Mixed Silk and Cotton

Throughout the period under review, it seems that, mixed fabrics and cotton goods comprised the largest bulk of production. The mixed piece goods exported by the Company were mainly *allabanees, cuttanees, carridaries* or *choradarries, chucklaes, cherconnaes, cushtaes, doreas, elatches, ginghams, jamdanees, nehallewars, nillaes, peniascoes, sooses, seersuckers*, and *mandilla*. Peter Mundy, in the thirties of the seventeenth century, referred to the production of *linen* striped with white silk.[10] Alexander Hamilton mentions a sort of 'tough grass' out of which *ginghams*, and *fiascos* were made at Balasore.[11] Further, he referred to the production of *rumals, qurrahs* and *lungis* of silk and cotton mixed at Balasore.[12] Sometimes, the filaments for silk manufacture were drawn from the barks of certain very white and very beautiful trees.[13]

Cotton Piece Goods (Plain or Painted)

But it was cotton piece goods, which numerically far surpassed other piece goods, whether of silk or mixed varieties in the production list of Orissa during the period under survey. The different varieties of cotton cloth manufactured in Orissa comprised of *sanas* or *sannoes, gurrahs, cassaes* or *khassa, ginghams, hummanees, orammalls* or *rumal, mulmul* or *malma, neelaes* or *nillees* or *nila, dimities, battlilies, charconnas, ambaries, seer bands, lungies, do-suties* and *salampores* etc. The quality of the different types of cotton piece goods in different areas varied widely as did their prices.

As regards the changes in consumer tastes, the 'Indian Craze' got set in about the 1680s and was a marked feature of the last decade of the seventeenth century. It is unnecessary to describe this trend in fashion, but it obviously acted as an active economic factor. The nature and extent of this fashion is revealed by J. Cary's Pamphlet of 1695 which states:

> "... It was scarce thought about twenty years since that we should ever see Calicoes, the Ornaments of our greatest Gallants (for such they are, whether we call them Muslin's, shades or anything else) when they were then rarely used, save in Shrouds for the Dead, and Chiefly among the poor who could not go to the price of finer Linnen, and yet were unwilling to imitate the Rich; but now few of them dress well till they are made up in Calicoes, both men and women calico shirts, Neck cloths, Cuffs, Pocket-Handkerchiefs for the former, Headdresses, Night-royals, Hoods, Sleeves, Aprons, Gowns, Petticoats and what not for the latter, besides India stockings for both the sexes".[14]

In Orissa, however, we see that during our period among the textile products, cotton goods held the important and foremost place. Peter Mundy (AD 1628–1634) furnishes a list of cotton textiles produced in Orissa. According to him, the following varieties were produced in the thirties of the seventeenth century, e.g., *cassaes* (fine and thin cloth), *ambarees* (canopied howdah), *chara cannaes* (chequered *muslins*, linen striped with white silk) *hamam* (a thick stout cloth used for wrappers) linen.[15] Some have the cheap, but well-made cotton manufactures were available at Balasore. In 1644, as recorded by William Foster the Surat factory was about to invest money in "*ginghams, sannoes, cassaes* and *hummanees* proper for England, intended to be sent unto you as a testimony of what cheap and well-made cloth those parts afford".[16] It is evident from the account of Lockyer, that "Balasore *ginghams*" had taken a prominent place in the field of textiles of Orissa at the close of the seventeenth century.[17] During the visit of Alexander Hamilton, Balasore had rightly become the emporium of textile products. The native weavers manufactured cotton cloths, *sannoes, cassaes, dimities, mulmuls, rumals, gurrahs, lungis, ginghams* and *spinascoe*.[18]

Sannoes or *sanas* was a kind of fine white cotton goods (Sanu). 'Balasore is a place celebrated for trade in fine white *calico* called *sanas*'.[19] Sir R.C. Temple thought that the *sanah* of the seventeenth century exists as *salu*, a cheap fine cloth, generally red, in common use in Orissa and Bengal.[20] But this seems improbable, as the one is white and the other red. In Hobson-Jobson, *sannoes* has been mentioned in the list of piece goods without any explanation of the term.[21]

Ginghams were defined as an Indian cotton cloth. The term probably of Indo-European origin denoted "a stuff made of cotton yarn dyed before being woven".[22] The Orissan variety of *ginghams* was superior to the Bengal variety available at Hugli, as Bridgman, sending a bale of the latter to the Company in AD 1650, did not recommend it. Its important feature was that it was woven with double threaded warps and wefts having a distinct texture.[23]

Balasore was an emporium of cotton yarn, cotton and *tassar* manufacturers of the interior hinterland and surrounding places.[24] Most prominent among the centres, arranged in order of quality of goods manufactured were Suro (Soro)[25], Harrapore (Harishpur),[26] Mohanpore (Muhunpur)[27] all specialising in the manufacture of *sannoes*. Of the varieties of cotton cloths produced at Balasore, *sannoes, dimities* and *mulmuls* – were noteworthy.[28] The factory records reveal that the English had a trade in *sannoes* produced in Balasore in 1661.[29] It has been observed that there were ample opportunities to explore the "very great straggling town" Balasore.[30] There were houses of a large number of merchants who had trade links with the European Companies. At a consultation the English factors at Balasore (16th December AD 1676) proposed to deal with the weavers of Soro for *sannoes*. Clavell writes in his account of Balasore that the weavers of Suro used to live in the "skirts of Balasore". Sir R.C. Temple, the editor of the *Diaries of Streynsham Master*, expresses doubt at this, saying that Suro can hardly be called outskirts of Balasore.[31]

In Balasore, coloured *ginghams, nillaes* and non-silk *rumals* were manufactured and sold to the East India Company in AD 1677. In AD 1677–78, St. George received 10,000 pieces of *ginghams* and 8,000 pieces of *sannoes* 10,000 pieces of *nillaes* from Balasore. According to the instructions of the Company, coloured *ginghams* of the finest sort 15,000 pieces *nillaes,* 18,000 pieces *sannoes,* 25,000 pieces and one hundred bales of cotton yarn, not hard spun or cross-reeled, were to be provided at Balasore.[32] In AD 1682, 3,000 pieces of *rumals* were to be provided to the English Company in Balasore.[33] *Dimties* were also produced in places around Balasore. These were white and have a very fine quality. The Company was to procure in 1698, 15,000 pieces of *dimities.*[34]

There was regular supply of Orissan cloth goods from Balasore to Patna in the 17th century. In the thirties of the seventeenth century, Peter Mundy noted that piece goods like *cassaes, ambarees* and *hammam* of Orissa were available at Patna in Bihar.[35] Forty years later, Thomas Bowrey observed that the various cotton manufactures of Balasore like *sanas, ginghams, orammalls* and cotton yarn and other goods were sent to Patna.[36]

Pipli, was known as a town for its cotton manufactures and it 'produces the same commodities that Balasore does'. Hamilton observed that Pipli was an important centre of trade and it produced broad cloths and fine cloths of superior quality of different varieties like *sannoes, dimities* and *mulmul.*[37]

Hariharpur, a village adjoining the modern Jagatsinghpur situated about twenty-five miles from Cuttack[38] was an important centre of trade and manufacture of all kinds of clothes.[39] In AD 1633, after the English settlement, Mr Colley purchased nearly 4,000 pieces of cloths at Hariharpur.[40] The Company's agent at Hariharpur wrote on 17th July AD 1633, to Cartwright at Balasore: "I can get musters of *cassaes* which are now making ..." at Hariharpur.[41] In AD 1642, *cassaes* and *sannoes* were available there for export.[42] Bowrey mentions that a considerable quantity of *calico* manufactured at Hariharpur were sold to the Dutch and English traders. He observes, that the "English and the Dutch are first to bring over land the textiles of Hariharpur to their respective factories at Balasore.[43] Walter Clavell, the English factor observed in AD 1676, that 'the cloth of Harrapore' could be brought to Balasore easily by the land route.[44]

The cotton manufactures of Puri were in a flourishing state in the thirties of the seventeenth century. Bruton mentions that *taffataes* and other cotton manufactures were produced in abundance. Ralph Cartwright, the English factor, proposed in 1633 to send persons to Puri to procure cloth from that place. Towards the end of the century, it continued to be a cloth-producing centre of Orissa. Hamilton in AD 1708, has referred, to abundant production of cloth at Puri.[45]

Jaleswar was an important centre of cotton production in the early part of the seventeenth century and continued with the manufacture of cotton cloth in the 70s of the 17th century as observed by Manrique.[46] Master, during his stay at Jaleswar obtained cotton on behalf of the East India Company.[47]

Cuttack, in the early part of the seventeenth century was a great centre of trade and all types of cotton cloth, both coarse and fine, were manufactured and available at a cheap price. In 1708, Hamilton mentioned, the *seerbands* and *sannoes* of Cuttack were cheaper than that of Balasore.[48]

Bhadrak and Jaipur produced *sahana, barbara, dosnti* and *tahti* varieties of cloth. Bhadrak was a town of about 1,000 houses and weaving was the sole occupation of the inhabitants.

William Bruton, the English factor, found Balikada, eleven miles south east of Hariharpur in AD 1633 a centre for production of colour fashioned cloth, which would have included cotton cloth.[49]

Ganjam was an important centre for cotton textiles in Orissa. Production of many kinds of calico, especially *battililes*, was evident from the account of Bowrey in the seventies of the seventeenth century at the coast of Gingelly, which extended from the river Godavari to Puri.[50] *Cassaes, mulmuls* and *sannoes* were manufactured in Ganjam during the seventeenth century. The textile of this part of Orissa was famous for its quality as well as cheapness of price.[51]

Balaramgarhi was also one of the important cloth producing areas of Orissa, situated nine miles east of Balasore. The East India Company had established their subordinate factory at this place. The English Factory Records mention frequently the textiles produced at Balaramgarhi.[52]

II

The principal characteristic of the export trade was that the merchandise was procured through the advance payment system (*dadni*). The weavers received them from the merchants. This was done much before the shipping season. This was in fact, the general structure of Indian maritime trade. Generally, the *dalals* or brokers established the link between the *paikar* and the merchant. Normally, *paikari, dalali* and the sea trade were considered to be three distinct professions. But sometimes influential merchants like Khem Chand would perform the all the three single-handedly. For the European, this was not possible because they neither could establish any communication with the primary producers nor were familiar with the market conditions and systems in their factories, where the prices of both the export and import commodities were mutually agreed upon for the goods to be disposed.

Orissa produced both agricultural and non-agricultural goods much in excess of her needs. The surplus was exported to different parts of India as well as other countries in Asia and Europe. The internal trade was carried on by both land and river routes; the coastal and external trade was carried on only by waterways.

As in most of the regions of India, textiles constituted a significant item of export from Orissa; but it was not on the same level as in Gujarat, Coromandel and Bengal, except in the case of certain varieties. The textiles exported from Orissa were of certain varieties, like cotton, silk and mixed-piece goods. *Tussar* was the most important among silk cloth. Though in quantitative terms the manufacture of *tussar* cloth was not so important, its quality was considered to be the best in Orissa.[53] The piece goods enjoyed predominance in the export of the Europeans throughout seventeenth and eighteenth century.

But the manufacture of cotton piece goods was widespread in the entire region of the province. The variety could be classified as coarse, fine and very fine. *Muslins* were fines in quality than the *calico*. The distinction between them is one of degree and each can also be further classified as coarse and fine. According to Moreland, in some cases, it is also difficult to say whether a particular textile should be classed as a fine calico or coarse *muslin*.[54] Generally, *muslins* were thinner in texture and lighter in weight than *calico* and they were suitable for wearing in hot countries.[55] The manufacture and trade of very fine *muslins* in Orissa were extremely limited as Dacca enjoyed an unquestionable supremacy in this field in the entire seventeenth and eighteenth centuries. But the other varieties of *muslin* and *calico* produced in the province were good and fit for export to both Asian and European countries. Some of the fine *muslin* produced in Orissa was *khaas, mulmul* and *adathi*. We do not know how far these varieties found a market in Asian countries, but the European procurement was not regular from Orissa and were mostly procured from Dacca, Malda, Santipore and Hugli areas, though in the eighteenth century, we find the *mulmul* of Kasijurah of Midnapore on the regular export list of the English.[56] The reason for such low procurement of these fine *muslins* from Orissa was that they were not as fine as the Bengali varieties. Peter Mundy also informs us that the *khasa* of Orissa available at Patna was thicker than the *khasa* of Sonargaon.[57] Generally all the *muslins* were embroidered either with gold or silver yarn or cotton or coloured silver yarn according to their variety.[58]

But the *calico* of both finer and coarser varieties dominated the export list with the industry also being widespread. The fine varieties of *calico* were manufactured in the Bengal's portion of Orissa extending from Puri to Midnapore districts. The areas comprising Gingelly coast produced the typical Coromandel variety of cloths. The main varieties were *salampore* and long cloth, which were woven at all the centres on the coast. These varieties had very good market both in Asian and European countries. A variety called *betilles* equivalent to Bengal *khasa* was woven in Srikakulam area, which had red stripes on the border.[59] In the early seventeenth century, this variety is said to have good demand at Batavia.[60]

Thus, all fine variety of *calico* of the above description were the major exports among the cotton textiles of Orissa. These fine varieties of *calico* may also be referred to as plain or ordinary *muslin*. Among the coarse kinds of *calico* mention may be made of *dessies, qurrahs, ambaris* and *lungis,* which were produced in the Bengal portion of Orissa. These were exported both to the Asian countries and Europe, though the first one was considerably more important due low prices. In the district of Ganjam, a long-cloth of coarse variety which the Dutch called Guinea or Negro cloth was manufactured. Another gunny twin and thread were also produced in the villages in this district.[61] From the mid to the later seventeenth century, the textile trade in Balasore and Pipli reached a high watermark with increase in their exports to the Asian and European countries.

In the first half of the seventeenth century, the cotton goods in India, did not feature so prominently in Western Europe and until AD 1660, there was practically no demand for *muslin* or prints as apparel and such dress-goods that were carried westwards were destined almost entirely for Africa or America. The trade network that had been established by the Portuguese could sufficiently meet the need of European household for tablecloths, napkins or towel through the sale of Indian *colico.*[62] However, when the English and the Dutch entered the commercial area, they realised the prospects of textile trade. Moreover, as England did not produce linen on a large scale, imports of calico did not face serious competition with the home industry. Their sale abroad brought money into the county and was therefore regarded with favour by the prevailing mercantilist opinion.[63] Previously, England also imported textiles, mainly silk, from Italy and France. But the Indian textiles, particularly the *muslin, calico,* silk or mulberry and *tussar* cloth of Bengal and Orissa proved to be much more competitive in England because of their low prices, though these silk-piece goods were not comparable with the former in quality.[64]

In the seventies of the seventeenth century, the English investment in Bengal expanded rapidly with the establishment of two more factories in Dacca and Malda. Among the piece goods regularly ordered by the court of Directors, mention may be made of *ginghams, sannoes, nillaes, mulmuls* and *rumals.* Of these, the former three varieties were favoured regularly from Balasore throughout the period under study.[65] In the early eighties of the seventeenth century, the English investment in all the factories in Bengal increased very rapidly. Balasore occupied the second largest place in terms of investment. In AD 1681, 72,500 piece goods were ordered apart from 20 bales, from this place. The order increased to 162,000 pieces plus 16 bales in AD 1682 and 158,000 pieces plus 16 bales in AD 1683.[66] The order for all the celebrated varieties like *sannoes, nillaes* and *ginghams* showed an increase as compared to the previous decade. The order for *ginghams* was 15,000 and 30,000 respectively and *nillaes* 18,000 and 36,000 respectively.[67] The order for *rumals* also increased, as in AD 1681 it was 7,500 pieces and in AD 1682, 20,000 pieces.[68] The share of Balasore in these two years was 40 per cent of the total *rumal* ordered from Bengal.[69] After AD 1683, the quantity ordered for Balasore factory gradually declined; for example, in AD 1684, the order for *sonnoes, ginghams* and *nillaes* were 20,000, 15,000 and 30,000, respectively.[70]

The Dutch trade in this decade was marked by a gradual decline with Japan and expansion of exports to Amsterdam. The frequent data regarding the export of mixed piece goods from Orissa are available to us. Comparatively speaking, the export of these mixed-piece goods were more than the cotton-piece goods throughout the period.[71] The Dutch also procured some cotton-piece goods from Orissa, which were not only exported to Europe, but also to some other countries in Asia. For example, in AD 1675, cotton *rumals* of Pipli were reported to be very useful in the trade with Spain and Turkey.[72] In the last decade of the seventeenth century, the procurement of some mixed piece goods like *ginghams* and *rumals* suffered due to the rebellion of Sobha Singh of Midnapore. But the end of the revolt showed a significant increase in the quantity of Bengal textiles exported to Batavia for Holland in AD 1698–1699.[73] However, there were some factions in the British Parliament who clamoured for the prohibition of Indian imports, and after some Act directed particularly against the silk or *tussar* piece goods of Bengal, (though it also made a reference to the manufacture of Persia, China and East India) the silk goods as well as mixed ones lost their predominance in the East India Company's export from the beginning of the eighteenth century.

Turning to the eighteenth century, it was a period marked by stagnation for Orissa as well as Balasore's trade with South-East Asia and Ceylon. Though we have no detailed data to compute the textile trade with these regions, it can be said that her textile trade carried on by Asian merchants with these regions came to an end. Textiles being the main item of export to South East Asia, the decline of shipping was a blow to the export of this commodity. However, the export of some ordinary piece goods to Maldives increased as a result of frequent shipping. For example, it increased from 1,200 pieces in AD 1680–81 to 2,600 pieces in AD 1699–1700.[74] After an initial setback in 1700, the export picked up and showed an upward curve from AD 1709 to 1720.[75] The Act of 1720 however did not stop the import of piece goods as they were allowed to come to England on condition of being re-exported. As a result, Amsterdam and Rotterdam became the chief markets for these goods especially, Indian wrought silk, Bengal mixed stuffs and *calico* painted, dyed, printed or stained in those ports.[76] But the position of Balasore as the investment centre had become insignificant in Orissa. The following table gives an idea of the volume of cloth trade of the Company in different parts of Bengal. The Calcutta Council ordered the following bales of clothes with different Factories in AD 1730 and 1731.[77]

Table 1

Factory	*Bales in the year*	
	1730	*1731*
Balasore	82	74
Calcutta	4,157	53,761
Dacca	544	200
Jagdea	7,716	7,900
Kasimbazar	2,681	2,000
Patna	252	250

From the above table, it becomes clear, that unlike the seventeenth century, the position of Balasore was downgraded in the eighteenth century. This was largely due to loss of importance of her mixed-piece and rising demand of fine *muslins* of Bengal.

The Gingelly coast was also important in period under study mainly for its grain trade and some sort of typical varieties of Coromandel textiles. As a textile-exporting centre, this region came to the forefront only in the eighteenth century. In this period, the main textile exporting centres for the Europeans were Vizagapatanam, Ganjam and Bimlipatnam. It is not possible to speak in detail about the trade carried by the Asian merchants. But all the varieties had a very good market in South East Asia. The Portuguese and the Dutch exported the coarse long cloth which was mainly available in Bimlipatnam and Ganjam areas to West Africa.[78]

The East India Company mainly collected *salampore*, long cloth and *battileis* from the Ginglly coast for Europe. The English used to bring its textile to Vizagapatanam from where they were carried to Fort St. George. It is not possible to give a detailed idea about each variety of textile. But it can be said that from the first decade of eighteenth century, the East India Company was keen to exploit the full potential of its business here. Of course, between AD 1724 and 1733, there was a general decline in the investment of the English on the Coromandel Coast. But this decline was felt particularly in Madras and Fort St. David and during this time, the northern region made up in some measure for the falling of the south. Again the trade in the years AD 1727 and 1729 were an exception.[79] However, lastly we can say that the textile trade of the Gingelly coast continued to prosper while the Bengal portion of Orissa showed a decline in the eighteenth century.

Notes

1. N.G. Mukherji, *Monograph on Carpet Weaving in Bengal*. Calcutta: Government Press, 1907, p. 1.
2. S. Chaudhuri, *Trade and Commercial Organisation in Bengal, 1650–1720*. Calcutta: Firma K.L. Mukhopadhyay, 1975, p. 192.
3. Strensham Master, *The Diaries of Strensham Master, 1675–1680*. R.C. Temple (ed.), Vol. I, London, 1924, p. 308 (hereafter Master's Diary).
4. S.A. Khan, *East India Trade in the XVIII Century*. Delhi, 1924, pp. 160–61.
5. Master's Dairy, Vol. I, p. 308.
6. A. Hamilton, *A New Account of the East Indies*. Vol. I, London: Argonaut Press, 1930, p. 217.
7. S. Chaudhuri, op. cit., p. 193.
8. William Bruton, "Account of Cuttack and Puri", *OHRJ*, 1961, No. 3, p. 48.
9. Sebastian Manrique, *Travels of Fray Sebastian Manrique*. C.E. Luard and H. Hoston (eds.), Vol. II, Oxford: Hakluyt Society, 1929, p. 99.
10. Peter Mundy, *Travels of Peter Mundy in Europe and Asia, 1608–67*. R.C. Temple (ed.), Vol. II, Nendeln/Liechtenstein: Haklyut Society, 1967, p. 155.
11. A. Hamilton, op. cit., Vol. I, p. 217.
12. Ibid.
13. H. Hosten, "The Abbate D. Matteo Ripa in Calcutta in 1709", *BPP*, 1914, Vol. 8, p. 54.
14. J. Cary, *Discourse on the East India Trade*. London, 1745, p. 4.
15. Peter Mundy, op. cit.; K.P. Mitra, "Discursive Notes", *IHC*, Vol. 1937, Lahore, p. 91; J.N. Sarkar, "Medieval Orissa's Sea-Ports: Balasore", *JBRS*, Vol. 36, 1950, p. 157.
16. William Foster (ed.), *English Factories in India, 1642–45*. London, 1915, p. 207 (hereafter E.F.I.).
17. Charles Lockyer, *Account of Trade in India*. London: Samuel Crouch, 1711, p. 302.
18. A. Hamilton, op. cit., Vol. I, p. 217.
19. T. Bowrey, *A Geographical Account of the Countries Round the Bay of Bengal, 1669–76*. R.C. Temple (ed.), London, 1905, p. 231n.
20. Ibid.

21. Hobson-Jobson, op. cit., p. 708n.
22. T. Bowrey, op. cit., p. 231n.
23. *Journal of Indian Textile History*, 1956, App. IX, p. 41.
24. J.N. Sarkar, "Medieval Orissa's …", *JBRS*, Vol. 36, 1950, p. 152.
25. Suro or Soro in Balasore district (Lat. 210160, Long 860490); J.N. Sarkar, *Studies in Mughal India*. Delhi, 1925, pp. 228–29.
26. Harrapore (Harishpur) near modern Jagatsinghpur, a town in the Mahanadi delta, about halfway between Cuttack and Harishpur Gar, a port at which the English landed.
27. A village previously in Orissa, now, in Midnapur district (Lat. 21050′, Long 87024′).
28. A. Hamilton, op. cit., Vol. I, p. 217.
29. EFI, 1661–1664, p. 149.
30. S.A. Khan (ed.), *John Marshall in India*. London: Oxford University Press, 1927, p. 7.
31. Master's Diary, Vol. I, p. 308.
32. Records from Fort St. George, Despatches from England, 1680–82, Madras, 1914, p. 67.
33. Records from Fort St. George, Letters to the Fort St. George, Vol. 1682, Madras, 1916, p. 63.
34. Records from Fort St. George, Despatches from England, 1696–99, Madras, 1929, p. 81.
35. Peter Mundy, op. cit., pp. 154–55.
36. Thomas Bowrey, op. cit., pp. 231–32.
37. A. Hamilton, op. cit., Vol. I, p. 215.
38. L.S.S. O'Malley, *Bihar and Orissa District Gazetteers*. Cuttack: Government of Orissa, 1933, p. 155.
39. William Bruton, op. cit., 1961, No. 3, p. 32.
40. J.N. Sarkar, "Medieval Orissa's …", *JBRS*, 1050, p. 153.
41. EFI, 1630–1633, p. 308.
42. William Hedges, *The Diary of William Hedges*. R. Berlow and H. Yule (eds.), Vol. II, London: Hakluyt Society, p. CLXXXII, 1887–89 (hereafter Hedge's Diary).
43. T. Bowrey, op. cit., p. 129.
44. Master's Diary, Vol. II, p. 84; EFI, 1642–45, p. XXVII.
45. A. Hamilton, op. cit., Vol. I, p. 214.
46. Sebastian Manrique, op. cit., Vol. II, p. 99.
47. Master's Diary, Vol. II, p. 86.
48. A. Hamilton, op. cit., Vol. I, p. 216.
49. William Bruton, op. cit., Vol. X, 1961, No. 3, p. 32.
50. Thomas Bowrey, op. cit., p. 122.

51. Records of Fort. St. George, Public Despatches to England, 1694–1696, Madras, 1919, p. 38.
52. L.S.S. O'Malley, B.D.G., Balasore, 1907, pp. 132–134.
53. Streynsham Master, *The Diaries of Streynsham Master, 1675–1680*. R.C. Temple (ed.), Vol. 2, London: John Murray, 1911, p. 85.
54. W.H. Moreland, *From Akbar to Aurangzeb*. London: Macmillan, 1923, p. 57.
55. Ibid.
56. S. Bhattacharya, *The East India Company and the Economy of Bengal, 1704–1740*. Calcutta: Firma K.L. Mukhopadhyay, 1969, p. 159. K.K. Datta, *Alivardi and His Times*. Calcutta: The University of Calcutta, 1939, p. 192.
57. Peter Mundy, op. cit.
58. Om Prakash, *Dutch East India Company and the Economy of Bengal, 1630–1720*. Princeton: Princeton University Press, 1985, pp. 61–52. Sushil Chaudhuri, "Textile Trade and Industry in Bengal Subah, 1650–1720", *IHR* 1(2), 1974, p. 264.
59. Om Prakash (ed.), *The Dutch Factories in India, 1617–1623*. Delhi: Munshiram Manoharlal, 1984, p. 185.
60. Ibid.
61. S. Arasaratnam, *Merchants, Companies and Commerce in the Coromandel Coast, 1650–1740*. Calcutta: Firma K.L. Mukhopadhyay, 1969, p. 49.
62. W.H. Moreland, op. cit., p. 123.
63. Ibid., p. 124.
64. Sushil Chaudhuri, op. cit., p. 265.
65. S. Chaudhuri, op. cit., pp. 260–62.
66. Despatches from England (hereafter DFE), 1680–82, p. 67; S. Chaudhuri, op. cit., 1(2), 1974, p. 268.
67. Ibid.
68. Ibid.
69. In 1682, 50,000 pieces were ordered in which Balasore's supply was 20,000. S. Chaudhuri, op. cit., pp. 263–64.
70. DFE, 1681–1686, pp. 108–09.
71. Om Prakash, Dutch East India Company (For detail), pp. 188–95.
72. Ibid., p. 60.
73. Ibid., pp. 213–14.
74. Ibid., p. 233.
75. Sushil Chaudhuri, op. cit., 1(2), 1974, p. 276.
76. S. Chaudhuri, op. cit., pp. 158–59.
77. Ibid., p. 164.
78. S. Arasaratnam, op. cit., pp. 49–99.
79. Ibid., p. 191.

11 Hydraulics and Water Management at Fatehpur Sikri[1]

Syed Ali Nadeem Rezavi

The remains of Fatehpur Sikri situated as it is on a ridge supply us some vital clues to the science and techniques of water management and hydraulics during the medieval period in north India. As a rule the pre-modern settlements generally grow around some major source of water, such as a river, lake or pond, which would not only act as a source of potable water, but also provide a barrier for security. Also it is generally observed that within a settlement, the areas closer to the source of water were reserved for the more influential sections of society. This may be seen at Delhi, Agra, Ahmedabad and Cambay. It is in this light that we may study the layout and water supply system at Fatehpur Sikri.

Before the founding of Fatehpur as a capital city, the village of Sikri was a "well-watered" ground which could sustain a large army. Babur testifies to the abundance of water at Sikri (on the plain by the side of the lake) when he mentions the Kol lake near this village.[2] When in AD 1571–72 orders were issued by Akbar to construct the city of Fatehpur, the major elements of the town were planned not around this lake, but on the broad top of the ridge and on the plain towards the south of the ridge. The scheme of placing the important structures and influential sections along the banks of the water reservoir was not followed at Fatehpur Sikri. Abul Fazl specifically mentions that the lake *(golābi)* was situated 'below the town' (*pīwast-i-shahr*) on whose banks "His Majesty constructed a spacious courtyard *(saffa)*, a *minār* and a *chaughāngāh* (polo ground), where elephant fights are organised".[3] Our sources however reveal that the lake of Fatehpur Sikri remained the major source of water supply to the city. Fr.Monserrate who visited the city in 1580–81 with the first Jesuit Mission says:

> To supply the city with water a tank has been carefully and laboriously constructed two miles long and half a mile wide. The work was performed, by the King's direction … .[4]

A survey of the dry bed of the lake in Sikri reveals that though it is formed by a natural depression of the ground between the Sikri ridge on the south and certain spurs in the north-west near the village of Rasulpur, it was regulated by the construction of two barrages, viz., the *Terah Mori* and the *Bāwan Mori.* The Terah Mori barrage, situated exactly to the north of the lake, on the Agra-Bharatpur road comprises thirteen arches which contained wooden sluices to release the excess water. To the north east of it is constructed the more heavily built barrage now popularly known as the Bawan Mori, which as the name suggests once comprised fifty two sluices. This earthen barrage forms the limit of the lake towards the old township of Nagar. The excess water released from these barrages flowed into a rivulet which then passed on towards the east, crossing the Agra-Sikri road near the town of Kirauli. This rivulet provided water to the area situated to the east of Fatehpur Sikri. The lake was fed through two sources, a channel drawing in the Utangan River, now known as Khari Nadi, and rain water. It appears that these barrages and dams which regulated the natural lake were constructed by Akbar in AD 1579. In a letter to Fr. Peres written in AD 1580, Fr. Henrique, a member of the Jesuit Mission reported:

> … about a year ago, in order to improve the city, water has been led in from somewhere to form a sizeable lake which is perennial. All the elephants, horses and cattle drink from it, and it also serves the teeming population for all purposes.[5]

Fr. Monserrate is however, much more explicit:

> Across to the end of a low-lying valley which was filled with water in the rains, (although the water afterwards drained away or dried up), a great dam was slowly built. By this means not only was a copious supply of water assured, but the discomfort of the climate was mitigated.[6]

It is interesting to note that our sources vary when they discuss the size of this lake, though all agree that it was very large. In 1580 In 1580 Fr. Monserrate, mentions it to be "two miles long and half a mile wide". In 1610, when William Finch visited Fatehpur Sikri, he found it "2 or 3 cos in length".[7] Nine years later, in 1619 when Jahangir ordered it to be measured, and found it's circumference to be 7 *koss*.[8] Later in 1633, Peter Mundy reported it to be "10 or 12 mile long".[9] Though these statements vary, they can partly be explained by the seasonal changes in the volume of water and partly by the vulnerability of the subjective estimation. But it is clear that at least until early years of Shahjahan's rule lake had maintained its extremely large size.

The demand for potable and irrigation water was also taken care of by digging a large number of wells (*chāh)* and constructing a number of *bāolis* (step wells) scattered all over the town. While interviewing a number of senior residents of Sikri, Nagar and Fatehpur during the course of my survey in the first week of March 1996, my informants spoke of the existence of as many as 52 *bāolis* and 108 masonry wells till the fourth decade of the present century. The survey of the area however revealed 14 step wells and over 60 wells, were all there when Fatehpur Sikri was built.

Among the step wells or *bāolis* which were located during the course of the survey, some were in a very dilapidated condition. Since these may soon disappear, it is worth recording them here: (1) Octagonal *bāoli* below the Hathipol; (2) Indārawali *bāoli* in the Indara Valley; (3) Shahjahani *bāoli* in the garden to the west of Hathipol Sarai; (4) a *bāoli* to the west of Babur's Bagh-i-Fath (only traces remain); (5) a *bāoli* near the so-called 'Matiya Mahal', near the Qush Khana (only outlines visible); (6) the four-storied grand *bāoli* around

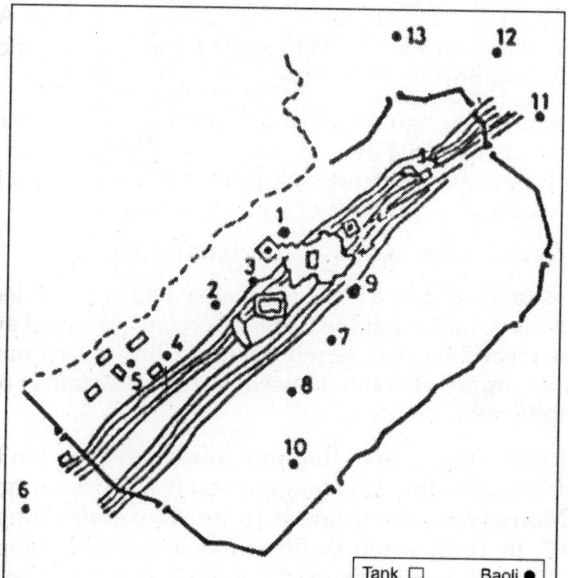

Sources of Water Supply

300 metres outside Ajmeri Darwaza; (7) Moti Bagh *bāoli,* now in the main bazar of Fatehpur, adjacent to the northern Gate of the four-laned Akbari *bazar,* popularly known as 'Pukhta Sarai'; (8) a *bāoli* in Muhalla Katra, near Purana *Dāk-khāna;* (9) Shah Quli's *bāoli* near the so-called 'Hakim's Baths' (the Imperial Baths); (10) a *bāoli* near the Gwalior Darwaza, adjoining a *Chahārbāgh;* (11) a *bāoli* around 300 metres outside Agra Darwaza, on the Agra road, (besides the modern wine-shop); (12) a *bāoli* to the east of the Agra-Bharatpur Road, outside the Lal Darwaza; (13) a *bāoli* to the north of the Bharatpur-Agra Road towards Nagar; and (14) a *bāoli* to the north-west of the Buland Darwaza, popularly known as the *jhālra.*

As can be seen from the appended plan depicting the placement of the sources of the water supply, these step wells were fairly evenly located all around the medieval settlement of Fatehpur Sikri. The survey further revealed that nearly all these *bāolis* (except Nos. 4 and 5) still contain water, and need only minor repairs to make them

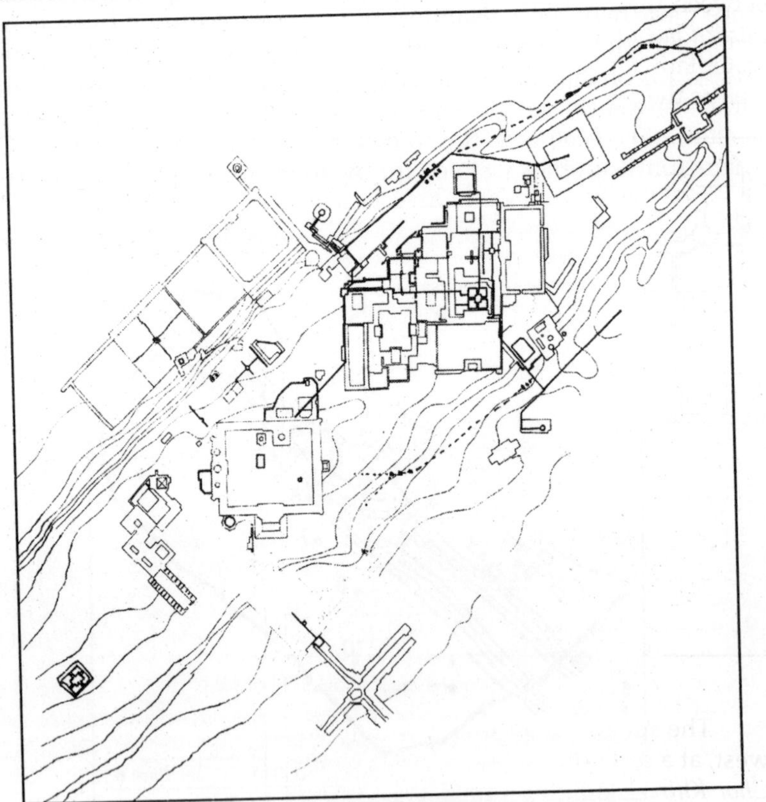

Distribution of Water

functional. Out of these 14 *bāolis* which could be located so far by the author, eleven belong to Akbar's period; one (No. 12) pre-dates the Mughals, to judge from the carving of the stone, probably a remnant of the Sikriwal Rajput rulers. Another belongs to Babur (No. 2). Baoli No. 3 was constructed during the reign of Shahjahan. Five of them (Nos. 1, 2, 7, 9, and 14) have octagonal wells, while the rest are square. Apart from being a regular source of potable water, these step wells also acted as pleasure resorts where the weary travellers could rest.

Apart from the major tanks in the palace complex, around six rectangular and one square tank *(hauz)* to store water were also located during the course of the survey. Of these, five were found in the so-called Indāra Valley, near the Ajmeri Darwaza, an area reserved for pleasure resorts and gardens. The sixth water tank is located in the residential area reserved for the nobility on the north eastern spurs of the ridge. This tank, which has recently been discovered, is situated between the nobles' quarters and the rows of excavated shops of the *Bāzār-i buzurg-i Sangīn* (the large stone market). This was a very large water tank to meet the needs of the houses of the nobles, the market, as well as the *sarāi* located near the Agra Gate.

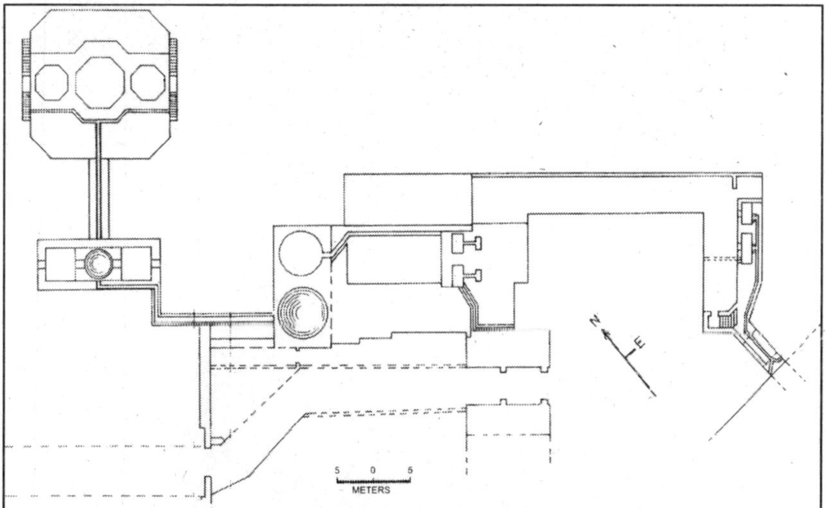

Northern Water Works (Hathi Pol)

The square water tank is situated atop the ridge on the extreme west, at a distance mid-way between the old Chishti quarters and the *Chor Khirki*. A modern cremation ground is situated adjacent to this tank. This tank resembles the *Anūptalāu*, in many of its features situated in the *Daulatkhāna*. Constructed with rubble stone, each side of this tank is provided with steps. In the centre is a square platform which is connected with a cause-way on all the four sides. It is interesting to note that all around this tank are the ruins of structures, which unfortunately have yet to be studied. The area also yielded a large number of low and medium quality blue ware sherds.

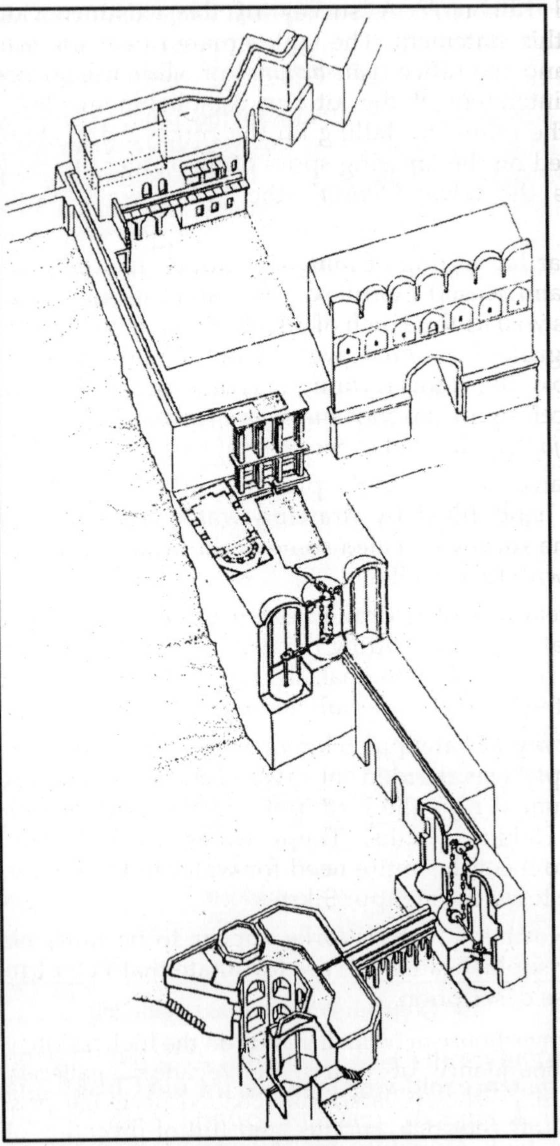

Northern Water Works: Elevation

Water harvesting appears to have been the major objective of the tanks situated on the top of the ridge. The major source of water for these tanks was rainwater which was harnessed into them through slopes and channels. We have the testimony of Abul Fazl that rainwater mixed with Ganges water was used to prepare food in the Royal Kitchen.[10] A survey of the existing water channels confirms this statement. The tank situated near the *matbakh* (Royal Kitchen) and the office *(yātishkhāna)* of Muhammad Baqir *Sufrachi*, the superintendent of the kitchen establishment was, and still is, filled by the rainwater falling on the courtyard of the *Daulatkhāna*. Constructed on the tapering spurs of the ridge, this tank, popularly known as the *'Hauz-i-Shīrīn'* (the sweet water tank), is raised on vaults.

Similar harvesting of rainwater appears to have been resorted to in the Jami Masjid Complex. The water falling on the floor of the Mosque as well as the Tomb of Shaikh Salim Chishti was collected in an underground covered water-tank *(birka)* constructed by the walling up of the vaults forming the courtyard of this mosque. Water from the ceiling of the mosque was directed to the octagonal *bāoli* and tank (*jhālra)* situated to the west of the Buland Darwaza.

The tanks situated in the plain (e.g., the Indara Valley) were, on the other hand filled by drawing water from the wells situated nearby. The survey revealed some of the water channels connecting the tanks with their wells.

The real feat of hydraulic engineering is however revealed in the elaborate system of lifting the water from the ground level to the top of the ridge where the palace complex, the nobles' quarters and houses of the main civic population were located.

A survey of Fatehpur Sikri reveals that the imperial system of water supply was divided into two sections – the Northern Water Works, situated near the Hathipol and the Southern Water Works, near the Hakim's Baths. These water works were apparently designed to meet the entire need for water in the palace complex as well as the town of Fatehpur Sikri.

The Northern Water Works appear to be more elaborate and technically sophisticated. We are fortunate that Peter Mundy gives a very precise description:

> The kings' house or Moholl stands on the highest hill, within which are aboundance of courts, conveyances, galleries, chowtrees *(chabutra)*, arches, pillars, tanks, chaboochaees (chahbacha), private rooms, all very rich, curious, and full of invention of painteinge, carvinage, etts; also a little garden. *The water* to *water it is also* to *fill*

> *the tanks alofte, and for their use is drawn from the valley, first into one tank and then from that into another higher, and so into* 4 *or* 5 *untill it came aloft, by that which wee in Spaine call Noraies.*[11]

At the first stage, the subterraneous water was raised through the construction of an octagonal *bāoli* below the Hathipol. This *bāoli* is shaped in the form of an irregular octagon, a chamfered square, with each of the principal sides measuring 15.45 metres. The chamfered sides of the octagon have lengths of 4.60 m. In the middle of this structure is an octagonal well with each side 2.90 m in length. To the north and south of the well are placed two octagonal chambers, the raised vaulted ceilings of which are visible as octagonal platforms from above. A 0.23 m wide water channel carried the water from this *baoli* to an artificial well situated to its west. This water channel runs on top of a 12.15 m long and 2.55 m broad aqueduct.

The artificial well (*chāhbachāh* or *chāh-bachcha,* receiving water from another well) which acted as the second stage for lifting water is a rectangular structure with two vaulted chambers flanking the circular well on two sides. The well itself is 10.6 m deep and 3.96 m in diameter. From here a water channel (0.23m in width) took the water to a second storage well, which is again flanked by two vaulted chambers. Between the two storage wells, the water was carried to an approximate distance of 10.50 m. The water brought to this second storage tank was then lifted to a large rectangular tank situated towards the south. A water channel then emptied the water into another rectangular trough constructed at the level of the floor of the Hathipol. Until this stage the water was raised to an approximate height of 30 metres from the ground level.

Just above the water trough projecting spaces are constructed to raise the water further to the fourth stage atop the roof of the quadrangle to the east of the Hathipol. A set of aqueducts diverted the water from this roof to the two storage tanks constructed on the first floor of the structure flanking the second gateway after the Hathipol. From here the water was again raised to a fifth stage, and, through a channel running atop the gateway entered the *haramsara* complex through the north-western corner of the so-called "Birbal's House" quadrangle (the first Imperial Palace).

Another branch of the water channel took the water through the northern walls of the Hathipol *daftarkhāna* to the bureaucratic establishments constructed below the *hauz-i shīrīn.*

A question which arises at this stage is, how the water was raised to the succession of the *chāhbachāhs.* We have already quoted Peter Mundy wherein he mentions the use of 'Noraies'. Bishop

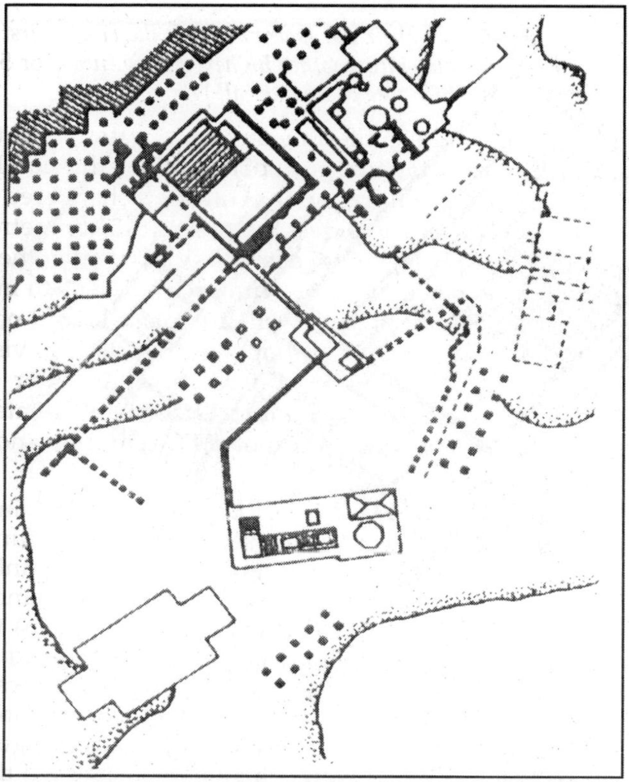

Southern Water Works (Hakim's Baoli & Quarters)

Heber visited Fatehpur Sikri in AD 1825. Describing the water works, although confusing its location and attributing it to be adjoining the Jami Masjid, he writes:

> ... and the whole hill on which the palace stands bears marks of terraces and gardens, to irrigate which an elaborate succession of wells, cisterns, and *wheels* appears to have been contrived adjoining the great mosque, and forcing up the water nearly to the height of its roof. The cisterns are still useful as receptacles *for* rain-water, but the *machinery* is long since gone to decay.[12]

Two *Akbarnama* paintings depict this water lifting device which was used to raise the water in the northern waterworks. (See plates 1 and 2) The first miniature, which was designed (or outlined; *tarah)* by Tulsi and painted (*amal*) by Bhawani, depicts a Persian wheel drawing water from a well near the Hathipol, which is under construction.[13] The second depicts Persian wheels at two stages near the Hathipol.[14]

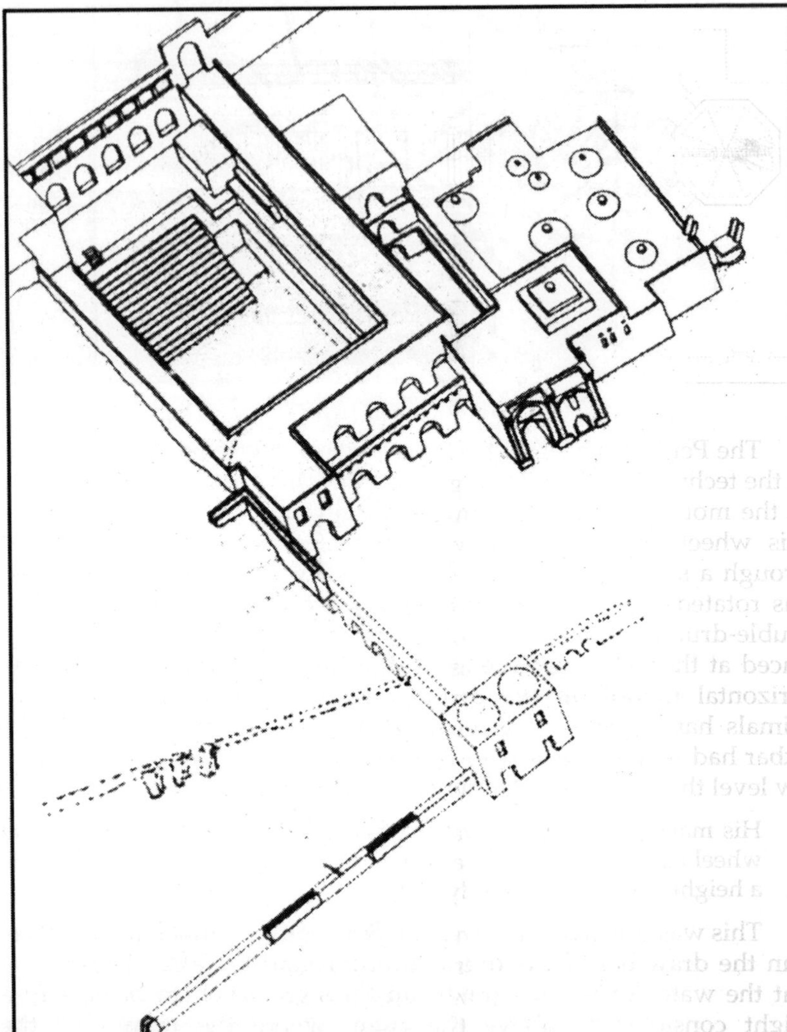

Southern Water Works (Hakim's Baoli & Quarters): Elevation

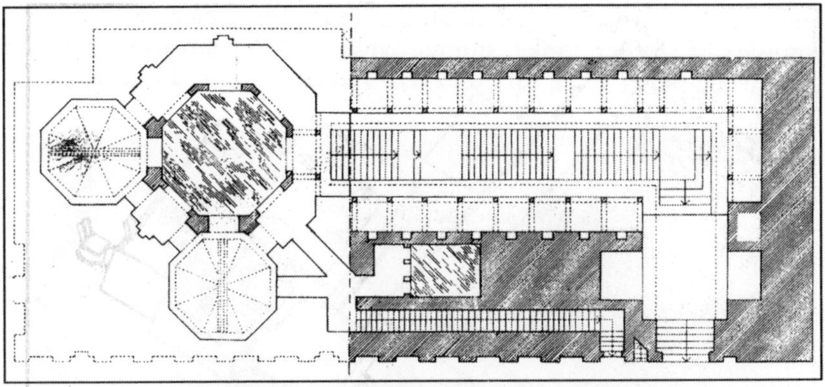

Shah Quli's Baoli (Hakim's Baoli): Plan

The Persian wheel, as we know, was a device which was based on the technique of pin drum gearing. It comprised of a wheel, fixed on the mouth of the well, which contained a string of pots *(māla)*. This wheel atop the well was connected with a second wheel through a shaft. The second wheel fitted with wooden pegs (pins) was rotated vertically by horizontally rotating a pin drum (i.e., a double-drum, the two layers of which are joined at the rim by pegs placed at the same distance as on the rim of the other wheel). The horizontal motion to the drum was provided through draught animals harnessed to it through a shaft. According to Abul Fazl Akbar had invented a way of raising water to a great height from a low level through the water, wheel. To quote:

> His majesty made such water wheels (*daulāb-ha*), and such (gear) wheel (*gardūn-ha*) were fixed thereon, that water may be carried to a height from distant low-lying places.[15]

This was due to his placing the pin drum at a much higher level than the draw bar of the oxen turning round the axle. This meant that the water would be drawn up through the chain of pots to a height considerably above the oxen, where the mouth of the receiving aqueduct could now be placed. It happens that two paintings in Nizami's *Khamsa*[16] prepared in Akbar's atelier illustrate such a device.

A closer took at the octagonal *bāoli* of the northern water works and its storage wells reveals the provision of stone shafts which once held the vertically rotating wheel of pots. The draw-bars which rotated this wheel were placed in the two vaulted chambers, which we have noted, flanked the well in the *bāoli* and the two storage wells. Similarly, protruding stone beams fixed atop the rectangular cavities near the Hathipol lifted the water atop the roof of the eastern quadrangle. Thus, we see that the northern waterworks was a complex of storage tanks, storage wells and Persian wheels which helped the water to be raised at five stages to reach the level of the Imperial complex atop the ridge.

Plate 1: "The Construction of Fatehpur Sikri", *Akbarnāma*, MS. V & A Museum, London, IS 1–1896, *f.* 86/117

The southern waterworks which centred on the *bāoli* attributed to Shah Quli, near the Hakim's Baths, survive in a much more dilapidated condition. From here the water was supplied not only to the *daulatkhāna* but also to the civic population living in the areas south of the ridge.

Shah Quli's *bāoli* appears to be the largest step well at Fatehpur Sikri and comprises of three storeys surrounding the octagonal well. Water drawn from this step well with the cutlasses fixed in the two chambers adjoining the well, was carried through ducts to a raised storage tank situated in front of the Hakim's Baths. From here the Persian wheels lifted the water, which was then distributed through aqueducts into three directions, the north, east and west. In the north, the water emptied into a tank situated near the massive *'Sukh tāl'* adjacent to the Imperial Baths, from where it was taken to the Baths and garden of the *daulatkhāna*.

Plate 2: "Akbar returns to Fatehpur Sikri from Gujarat", *Akbarnāma*, Ms. V & A Museum, London, IS 1–1896, *f.* 111/117.

In the area between the southern wall of the *daftarkhāna* and Shah Quli's *bāoli* a number of piers which once carried water ducts still survive. Similar aqueduct appear to have been constructed to the east of the storage well. During the course of the survey, bases of two such piers were encountered.

The water thus supplied from the northern and southern water works to the Imperial complex was distributed to various sections through the conducts running between the wells. Unfortunately, we encounter a number of gaps in these channels running through the Imperial complex. This is probably due to the renovation work which was undertaken under Lord Mayo and Lord Curzon. A look at the appended map would however point out that still enough evidence survives to connect the various palace complexes. The map would also suggest that the major supply to the Imperial complex was through the northern waterworks, whereas the southern water works catered mostly to the needs of the civic population, a function which it still performs.

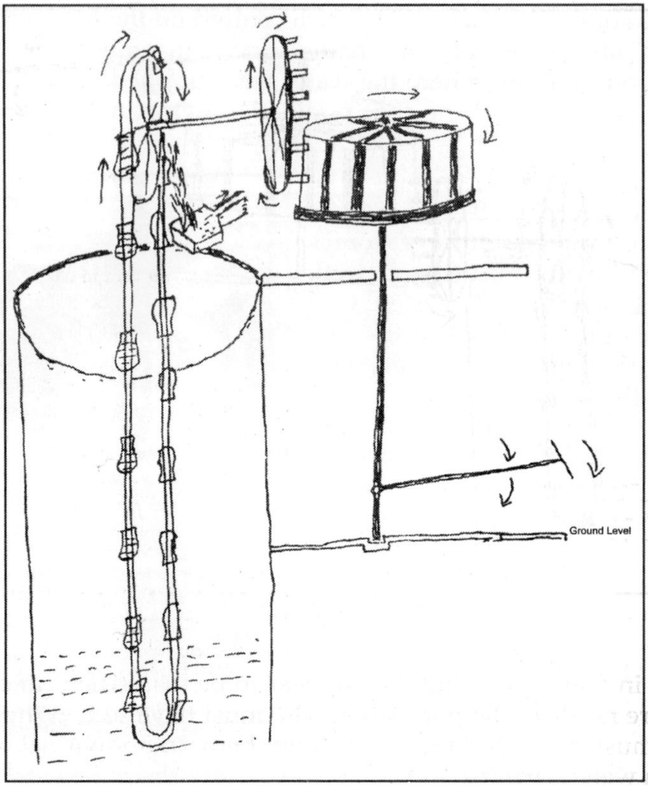

Persian Wheel as Depicted in Khamsa of Nizami

Apart from the *bāolis*, wells and tanks, a number of piers of the aqueducts survive around the city of Fatehpur which gives us some idea of how water was carried from one area to another. For example a series of such piers survive in the area below *hauz-i shīrīn*, on the slopes of the northern ridge, below the so-called Tansen's Baradari and the excavated residential structures. These piers carried the aqueducts which connected the northern works with the large water tank situated near the nobles' houses, as mentioned earlier. This massive tank is 28.10 m. wide and 67.40 m long. Probably the entire needs of the eastern area were met through the water stored in this tank.

It is also important to note that the individual residential structures, which have been excavated at Fatehpur Sikri in most cases, seem to have had their own water storage tanks.

There are still areas in Fatehpur Sikri, for example the areas in the south-west about which not much is known. A number of wells

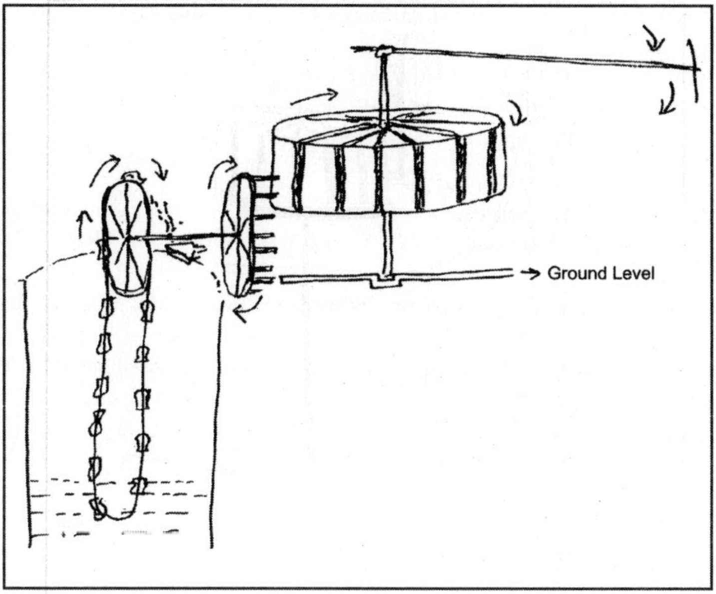

Persian Wheel

survive in that region, but they appear to be insufficient to cater to the entire needs of the population who must have settled there. The water must have been carried from here to individual houses through water-carriers.

Notes

1. This article was originally presented as a paper at a Seminar on Archaeology of Medieval India held by the Centre for Archaeological Studies and Training, Kolkata in 2001.
2. A.S. Beveridge (tr. and ed.), *Baburnama*. New Delhi, 1970, pp. 548–88.
3. Abul Fazl, *Ain-i Akbari*. Vol. II, Nawal Kishore, Lucknow, 1882, p. 84. Our surveys on the banks of the lake have also revealed only pleasure resorts, pavilions and gardens in the area between the ridge and the lake.
4. Fr. Monserrate, *The Commentary*. J.S. Hoyland (tr.) and S.N. Banerjee, Delhi: OUP, 1922, p. 31.
5. Letter from Fr. Henrique to Fr. Peres, *Letters from the Mughal Court: The First Jesuit Mission to Akbar (1580 –1583)*. J. Correia-Afonso (ed.), Bombay, 1980, pp. 22–23.
6. Fr. Monserrate, op. cit., p. 31.
7. William Finch, *Early Travels in India: 1583–1619*. W. Foster (ed.), Oxford: Oxford University Press, 1921, p. 150.

8. Saiyid Ahmad (ed.), *Tuzuk-Jahangiri*. Ghazipur and Aligarh, Delhi, 1864, p. 259.
9. Peter Mundy, *Travels in Europe and Asia 1608–1667*. R.C. Temple (ed.), Vol. II, London: Hakluyt Society, 1907–36, p. 230.
10. *Ain*, op. cit., I, p. 22.
11. Peter Mundy, op. cit., II, p. 228.
12. Bishop Heber, *Narrative of a Journey through the Upper Provinces of India from Calcutta to Bombay, 1824–25*. Vol. II, London: John Murray, 1828 p. 353.
13. 'The Construction of Fatehpur Sikri'. *Akbarnāma*. Ms. V & A Museum, London, IS 1–1896, *f.* 86/117.
14. 'Akbat returns to Fatehpur Sikri from Gujarat'. *Akbarnāma*. Ms. V & A Museum, London, IS 1–1896, *f.* 111/117.
15. *Ain*, op. cit., I, p. 199.
16. Ms. British Museum, Or. 12208, *ff.* 65 (a), 99 (b).

12

Air Dynamics and Building Construction in Medieval India

A Case Study of Sheesh Mahal of Farrukhnagar

S.S. Subodh

History is a discipline which links past with the present and present forms the basis for future.[1] It indicates the growth of human mind in which unique facts of life are collected/classified and interpreted in a scientific way. Archaeology is perceived as a discipline and also a method enabling historians to discern facets of human culture in the past that were not adequately recorded.[2] It not only includes the collection of old objects crafted by man, but also various methods to search for objects and the attempts to reconstruct the history of past events by the study of objects or things so collected. As a discipline, it is concerned with the theory and techniques required to draw conclusions about past human behaviour for material remains. "It attempts at how especially the monuments eloquently speak of the cultural movements and integration through the ages with some appropriate illustrations."[3]

Medieval past has been generally interpreted with a particular mindsetting. It has normally been understood to be a period of stagnation, bigotry and autocratic rule devoid of any scientific and technological advancement. "Some studies have glibly dubbed Indian society of the past as 'static': a fantasy of a people impervious to innovation and change."[4] Thus, medieval is perceived to be retrogressive, static or sluggish and, hence, devoid of any urge to create or even learn new techniques of production. However, this perception does not rest on a sound footing and in recent times a number of studies have tried to break the myth.[5]

Medieval archaeology is concerned with study of medieval archaeological relics.[6] The monuments of medieval India offer a subject of great attraction not only to the students of history of art and architecture but to the explorer of history of science and technology too. "The history of technology in medieval India began by largely concerning itself with the process of production and mechanical devices, that is, those sectors which had a direct bearing on the economy of medieval India. But, in the last two decades, the interest in the study of technology has broadened into new areas of investigation. However, one particular aspect of pre-modern Indian technology, namely, civil engineering, has not yet received the attention it deserves."[7] While several studies have been made on the capital cities, forts, palaces and other monuments of the Mughal emperors, the architectural remains left behind by the nobles, especially their residential houses have not received much attention. The manner in which some of the well-documented public buildings of the medieval period continue to be neglected also express our sense of history and how we view the past.[8] "Our contemporary sources record the zeal of numerous nobles for building lofty and beautiful mansions for themselves as well as for spending large amount of their income on public welfare."[9]

The present paper intends to make a study of the 'Sheesh Mahal' at Farrukhnagar. Sheesh Mahal does not become an object of study simply because of its relation with the past. It draws the attention of a historian for, its architectural technique demonstrate, how by steering the movement of air into the structure helped in keeping the interior cool even in the hot months of summer.[10] The technique becomes more appropriate when we take into account the fact that the climate of a culture area influenced the form of dwelling houses in the medieval era and in the "seventeenth century Mughal India, Bernier seems to have held a theory bearing on the climate-architecture connexion".[11]

The town of Farrukhnagar is situated in the midst of a sandy sterile tract to the north-west of the Gurgaon district in Haryana. During the Mughal emperor Farrukhsiar's reign, Dalel Khan was made governor of the district with the title of Faujdar Khan, by which he is popularly known. He farmed the Hansi-Hissar tract for Rs 1.5 crores and acquired extensive proprietary rights. In AD 1713, Faujdar Khan obtained a grant of land and laid the foundation of a fort and city, which he called Farrukhnagar, after the name of his master.[12] The Sheesh Mahal or Nawab's palace was begun by him and finished by his son, Kamgar Khan.[13]

Plate 1

Plate 2

Plate 3a

Plate 3b

Plate 3c

The structure under review gains its name from its glass decoration, some traces of which can be still seen.[14] It is in trabeate style and supported with a courtyard.[15] The materials used in the structure include red sandstone, bricks and wood. Despite being under the patronage of state archaeological department, it is in a dilapidated condition. At present, it houses the office of Gram Panchayat too. It is a north-facing structure[16] and can be divided into three parts, viz., central portion[17], right portion[18] and left portion.[19] The right portion of the structure is in ruins as is evident from the plates. Since the garden and the water supply system formed a basic part of housing complexes of the Mughal aristocracy,[20] the structure too contains a garden in the courtyard which has given way to barren land because of unkempt condition.[21] The right end of the complex also contains a well which was most probably used to water the garden of the complex and run the fountain in front of the main structure.[22] The main structure is of red sandstone with foliated arch or mihrab. It is 39 feet 7 inches long and 162 feet wide.[23] The height of the main structure is approximately 20 feet.[24] The entry to the structure is through a gate whose ceiling bears Khurasani arch and the decoration of the ceiling can still be seen.[25]

Plate 4

Plate 5

The central structure has a raised platform in front of it measuring 68 feet 7 inches by 21 feet. The height of the platform from ground is 2 feet 6 inches.[26] The platform contains a tank. The outer area of the tank measures 14 feet 9 inches by 12 feet 10 inches. The dimension of the inner area of the tank is 9 feet 4.5 inches by 7 feet .5 inches. The depth of the tank is 35 inches. There are two holes visible in the tank. The diameter of the upper hole is 7 cms and that of the lower is 9 cms.[27] In front of both sides of the platform there are remains of a small platform with foliated arch and a niche in it.[28] There are also recesses at four places at the bottom of the floor of the central structure.[29] The interior of the central structure is in a ruinous state.[30] It has a false ceiling of wood which could effectively curtail heat in summer months.[31]

Plate 6

Plate 7

Plate 8

Plate 9

Plate 10

Plate 11

Plate 12

Plate 13a

Plate 13b

Plate 14

Plate 15

Plate 16

Plate 17

The technique adopted for keeping the structure cool was very simple. As in Samarra (Iraq), the 9th century capital of the Abbasid caliph Mu'tasim, it has an underground chamber.[32] The wind catcher deflected the outside air current into a channel built in the wall.[33] The air was allowed to enter the complex from three sides. The wind entering from the front entered the channel through wind catchers erected at the front. Plate 18 shows the wind catcher erected at the front. Similarly, the wind entering from sides entered through wind catchers as shown in Plate 19 and entered the channel built in the wall as shown in Plate 20. The air thus directed entered the cellar and the central structure through passages in the wall on both sides as shown in Plate 21. The cold air passed through the rooms, the cellar and scattered to the courtyard, pushing warm air upwards. The warm air which is light in volume, when replaced by heavy cold air, escaped into the atmosphere through flues in the ceiling.[34] The warm air of the cellar escaped through these flues while flues in the ceiling as shown in Plate 22 released the warm air into the atmosphere from the rest of the structure.

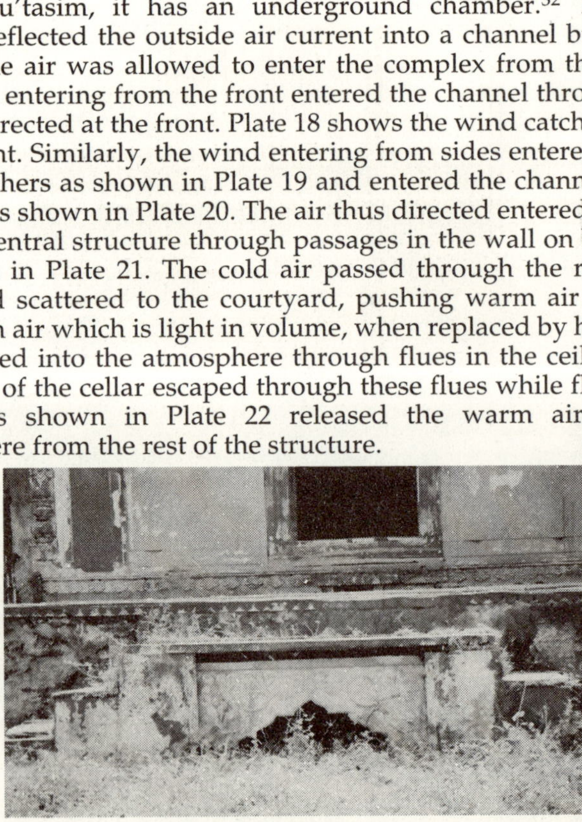

Plate 18

Plate 19a

Plate 19b

Plate 20

Plate 21

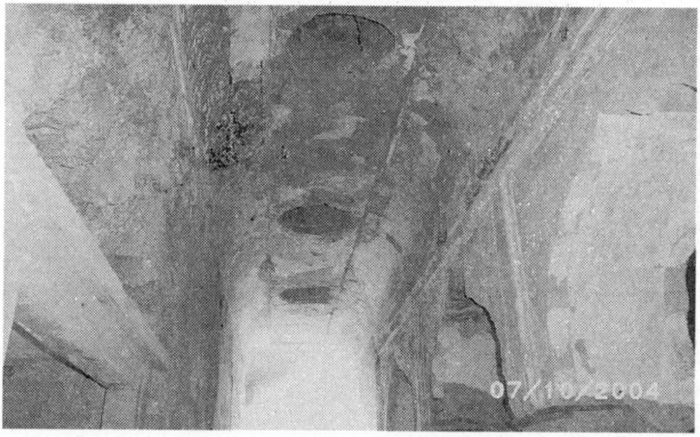

Plate 22

Thus, the study of structure's architectural design explains the simple method of keeping the structure cool by regulating the airflow. The problem of ventilating inner and underground chambers with the help of flues and chimneys was not unknown to the architects of Middle East and India.[35] In the areas of hot and humid climate like lower Sind and Afghanistan, we get information of construction of vertical structures known as 'badgir' to deflect the outside air current into the building to cool the whole interior.[36] The structure under review registers its place in the history because, it amply illustrates the usage of medieval scientific and technical knowledge and is in consonance with nature. "The production of any monumental building provides a matter of general interest, it deflects men's minds from the internal to the external, in which religious and other barriers are broken down, all differences become merged, in a unified effort of craftsmanship, so that humanity becomes one."[37] The study undertaken generates hope that if this method is put to use in modern times it would surely make the environmentalists happy.

Notes

1. E.H. Carr, *What is History?* London: Palgrave, 2nd edition, 1987, p. 108.
2. Iqtidar Alam Khan, 'Medieval Archaeology in South Asia: A Brief Review', *Proceedings of Indian History Congress*, 63rd session, 2002, p. 1284.
3. A. Sundara, 'Archaeology and National Integration', *Proceedings of Indian History Congress*, 45th session, Annamalai, 1984, p. 852.
4. A.J. Qaisar, *The Indian Response to European Technology and Culture*, Delhi: Oxford University Press, 1982. A.J. Qaisar further points out that a hypothesis has been offered that the pre-twelfth century India saw remarkable progress in science and technology but innovations were inhibited by a sense of insecurity generated by Muslim rule until the Europeans rescued us. Ibid., p. 2.
5. See A.J. Qaisar, *Building Construction in Mughal India*, Delhi: OUP, 1988; A.K. Bag (ed.), *History of Technology in India*, Delhi: Indian National Science Academy, 1977; Aniruddha Ray and S.K. Bagchi (eds.), *Technology in Ancient and Medieval India*, Delhi: Sudeep Prakashan, 1986; Debiprasad Chattopadhyaya (ed.), *Studies in the History of Science in India*. New Delhi: Editorial Enterprises, 1982; Irfan Habib, 'Medieval Technology: Exchanges between India and the Islamic World', *Aligarh Journal of Oriental Studies* 2, 1985; 'Joseph Needham and the History of Indian Technology', *Indian Journal of History of Science* 35.3, 2000; *A Shared Heritage: The Growth of Civilizations in India and Iran*, New Delhi: Tulika Books, 2002; Lynn White Jr., *Medieval Technology and Social Change, Oxford: Oxford University Press, 1962.*
6. R.N. Mehta, *Medieval Archaeology*, Delhi: Sterling Publications, 1979, p. 24.
7. Ravindra Kumar, 'Some Observations on Masonry Bridges in Mughal India', in A.J. Qaisar and S.P. Verma (eds.), *Art and Culture*, Jaipur: Publication Scheme, 1993, p. 143.
8. Sanjay Subodh and Amit Chaudhary, 'Material Remains and Historical Sense: A Study in Medieval Archaeology', *Proceedings of Indian History Congress*, 61st session, Kolkata, 2000–01, p. 527.
9. Mohammad Afzal Khan, 'The Mughal Elite: Their Building Activities and Cultural Values', in A.J. Qaisar and S.P. Verma (eds.), op. cit., p. 133.
10. We get reference of a number of structures where the basement coupled with hydrology helped in keeping the interior cool. Fatehpur Sikri, Red Fort at Delhi, Rai Mukund Das Chatta and Jal Mahal at Narnaul are some of the examples which strengthen the fact. Percy Brown while giving details of the architecture of Red Fort at Delhi states of a provision of a full and

continuous supply of water distributed throughout the entire enclosure. The water entered the fort through a sluice under the Shah Burj and was carried by channels "under and around the marble pavements of the royal pavilions, so contrived that each apartment included all the accompaniments of a water-palace". Percy Brown, *Indian Architecture,* Bombay: D.P. Taraporevala (sixth reprint) 1975, p. 104. In the 18th century Lucknow, the construction of Martin's palace too was based on the same principle and gave the Englishman relief from the scorching heat of Lucknow. "The basement storey comprises two caves or recesses within the banks of the river and level with its surface when at the lowest decrease. In these caves (Martin) generally lived in the hot-season, when the increase of the river obliged him to remove … . Within the basement apartments were baths and fountains which sprayed water against the windows, which, when the hot winds prevail during the spring months, are kept covered in the day time with frames filled with green brambles; these being kept constantly, wet by the fountains, cool the wind in its passage into the apartments, and thereby procure a constant temperature within, proportioned to the strength of the wind abroad, and capacity of the frames to retain the water dripping." Rosie Llewellyn-Jones, 'A Fatal Friendship: The Nawabs, the British and the City of Lucknow', pp. 135–37 in Abdul Halim Sharar, Rosie Llewellyn-Jones and Veena Talwar Oldenburg, *The Lucknow Omnibus,* New Delhi: Oxford University Press, 2001. However, apart from Sheesh Mahal, the author has also come across a structure in Hyderabad, where the Air Dynamics was used to keep the structure cool.

11. A.J. Qaisar, *Building Construction in Mughal India,* op. cit., pp. 1–2.
12. *District and State Gazetteers of the Undivided Punjab,* Vol. IV (Four Volumes Bound in Two), pp. 396–97. However, The Maathir-ul-Umara does not make mention of any official known as either, Dalel Khani or Faujdar Khan employed under Farrukhsiyar. See Shah Nawaz Khan, *The Maathir-ul-Umara,* Eng. Tr. H. Beveridge, 2 Vols., Delhi: Motilal Banarsidass, reprint 1999.
13. Ibid., see Plate 1.
14. See Plate 2.
15. See Plates 3a, b and c. The parapet, brackets, pillars, beam and the flat roof bear a testimony to the fact.
16. "The Mughal architects took much advantage out of the solar track. The fact that the sun rises in the east, travels through South and sets in the West was a guiding factor in the allocation of seasonal quarters. So, summer houses and palaces were in the north or in the West where the blazing sun will least trouble them." K.K. Muhammad, 'The House of the Nobility in Mughal India', *Islamic Culture,* Vol. LX, No. 3, July 1986, p. 86.
17. See Plate 4. The interior is divided into two storeys and is reflected in the facade by means of alcoves and arched bays. See Plates 14 and 15 for the division of the interior.

18. See Plate 5.
19. See Plate 6.
20. A.J. Qaisar, *Building Construction in Mughal India*, op. cit., p. 7.
21. See Plate 3.
22. See Plate 7.
23. The main structure contains a basement beneath it. At present it is not possible to enter the basement as entries to it are closed and it is full of bats. So keeping in view the measurement of the upper chamber, the basement too must be of approximately that size.
24. The author had no means to measure the actual height. The approximate height given is that of the ceiling from the floor of the structure.
25. See Plates 8 and 9.
26. See Plate 4.
27. See Plate 10. The tank was filled with water, lifted from the well, through channels. Most probably the lower hole was used to fill the tank and through upper hole the excess water was discharged which may have been used in the garden. Water supplied through another channel ran the fountain. The fountain gave a soothing effect in hot months of summer. A small structure in front of the tank probably functioned as the inlet of water for the fountain. The structure is 11 feet x 6 feet 9 inches. The actual height of the structure could not be measured as it is filled with earth. See Plate 11.
28. See Plate 12. However, the platform on the right side is in ruins. See Plate 1.
29. See Plates 13a and b.
30. See Plates 14, 15 and 16.
31. See Plate 17. The false ceiling ensured that the heat was trapped in the gap between the roof and the ceiling. Hence, the heat in much less degree was transmitted to the chamber below.
32. K.A.C. Creswell, *Early Islamic Architecture*, Britain, 1958, p. 285. The excavation at Champaner has also reported a similar underground cellar. See *Indian Archaeological Review*, 1978, p. 12.
33. See Plates 18, 19a and b and 20. The passage through which air was transported now houses bats so it was not possible for the author to photograph the openings from a close range.
34. See Plate 21.
35. See K.N. Chaudhuri, *Asia Before Europe: Economy and Civilization of the Indian Ocean from the Rise of Islam to 1750*. Cambridge: Cambridge University Press, 1990, p. 204.
36. George Michell (ed.), *Architecture of the Islamic World*, New York: Paperback, 1978, p. 203, *Cf.* K.K. Muhammed, op. cit., p. 94.
37. Percy Brown, *Indian Architecture*, Bombay: D.B. Taraporevala (sixth reprint) 1975, p. I.

13 Minting Technology in Mughal India

Najaf Haider

The medieval Indian exchange economy was based on the circulation of coined money to meet the transaction demands of the commercial and fiscal sectors and households. In order to ensure a high degree of uniformity and control over the monetary system, medieval Indian states generally permitted the circulation of imperial coins minted within their territorial jurisdictions. The Mughal state took special measures to standardise the circulating medium by minting currencies of uniform weight and fineness and restriking older issues into legal tender.[1] The success of the policy of giving currency only to money issued by the state in a large exchange economy was ensured by open coinage. The principle of open coinage entitled the suppliers of gold, silver and copper, mostly *sarrafs* (buyers for the mint-cum-money changers) and big merchants, to obtain freshly minted coins for the market on payment of mint charges.[2] Mints were located in market towns and capitals, and close to the sources of monetary metals, such as copper mines and port cities.[3] The efficiency with which the mints worked and the techniques they employed for coin production were, therefore, matters of critical importance for the performance of the monetary economy.

Mint organisation figured prominently in medieval administrative literature, in which their stated objective was to serve the people through a better organised currency system as well as to uphold the symbols of imperial authority. For Abul Fazl, a thoughtful commentator on Akbar's administrative measures, the successful working of the mint was linked to the prosperity of the Empire. The first book of his voluminous work contains chapters on the organisation of the mint (*dar uz zarb*) and the technology of coin production.[4] In one chapter, Abul Fazl offers a statistical account of the costs of manufacturing coins with a complete breakdown of the costs of labour and raw materials, seigniorage and the 'profit of the bullion merchants' in the last quarter of the sixteenth century.[5] Later sources supplement, and on some points modify, the description of the metropolitan mint given by Abul Fazl. In what is perhaps the only surviving mint record of the Mughal Empire, preserved in an administrative manual, we have another statistical account of a mid-seventeenth century provincial mint.[6] Information on the duties of mint officials and the process of coin production is also available in two early eighteenth century manuals of eastern India.[7] References to the working of the seventeenth century mints of Surat, Ahmadabad, Dacca, Rajmahal and the Deccan abound in the commercial correspondence of the English and Dutch East India Companies.

The Organisation and Production Capacity of the Mint

The mint was a centralised institution owned and operated by the state. As a fiscal unit, it fell under that branch of the department of finance (*diwani*) which dealt with the imperial revenue (*khalisa*).[8] In its routine functioning, the mint was under the direct control of the superintendent (*darogha*) and an imperial official of the same rank (*amin*).[9] The officials supervised the purchase of monetary metals and production of coins and managed the mint workmen. They were responsible for maintaining the shape, weight and fineness of coins in accordance with the guidelines issued by the centre.[10] Together with the treasurer (*ganjur*), the two officials were in charge of mint revenue as well.

Supervision was supported by accounting where the professional accountant (*mushrif*) maintained a systematic daybook (*roznamcha*) of the intake and production, and income and expenditure of the mint.[11] The record of the Patna mint, available only in fragments, bears eloquent testimony to the organised and detailed method of accounting observed in the mint.[12]

The mint functioned essentially as a craft centre for the mass production of coins of various metals, types and denominations.[13] The industrial organisation of the mint was based on a division of skill and labour across three broad sections, viz., assessment, refining and production. Each section had several departments in which artisans and labourers were engaged to perform specific tasks under the close supervision of the head craftsmen and mint officials, such as the master assayer (*sahib-i-aiyar*) who looked after the technical side of the operation.[14] Due to the nature and scale of its output, the economic significance of the mint far exceeded that of the other artisanal workshops which produced mainly for imperial and elite consumption.[15]

The size of the workforce employed in the mint is not known to us since the Mughal records describe the artisanal categories collectively and do not give sufficient indication of the total strength of each department. We know from the records of the Calcutta mint operated by the British in the last quarter of the eighteenth century that some departments could have more than one shop of head craftsmen and their teams performing the same task independently. For instance, there were 20 blank-cutters and 'each has a separate shop in which seven persons are employed'. In a lean season, the total workforce of the Calcutta mint, from the master assayer to the watchman, comprised more than 120 men.[16]

The yearly production of a mint was determined by the supply of metals, the technology and workforce employed in coining and operational costs. At the time of the year when merchant ships returned from the Red Sea and the Persian Gulf, there was a tremendous rush to coin money at the coastal mints, such as the one at Surat, which received foreign bullion.[17] When the season was over, the mints closed down and waited for the supplies to accumulate before commencing a fresh cycle of coining.[18] The availability of workmen all year round was another determinant. There was a premium on skilled labour and craftsmen were contracted on subsistence allowances when there was no work.[19] The workmen closed shops during certain days of the month (particularly on general holidays such as *ekadashi* and *amavas*) and also took leave to attend to other ritual obligations.[20] It was estimated in the late seventeenth century that the Surat mint worked an average of 20 days a month owing to the social customs and other preoccupations of the mint workers.[21] The mints were capable of raising their output if pressures mounted or the volume of supplies increased over a period of time sufficiently long to recruit additional workmen or expand the physical space in the mint building.[22]

Mint Prices and the Supply of Monetary Metals

The mint derived its supply of monetary metals mainly from internal stocks (demonetised coins and hoards) and foreign trade. The recoinage of internal stock was induced by weight loss (worn-out coins), changes in the type or weight of coins (Jahangir's recoinage of his heavier issues and Shahjahan's minting of all Nurjahani coins with his title), or the demonetisation of the coins of previous emperors (*padshahan-i sabiq*) and regional rulers.[23]

Due to a favourable balance of trade, India regularly imported foreign currencies, bullion and raw copper to replenish its internal stock.[24] In the last quarter of the sixteenth century, developments on a world scale had important consequences for India's monetary economy. The eastward migration of Spanish-American silver, together with the political unification of trade routes between the Levant and the Indian Ocean by the Ottomans and between the hinterland and coastal cities by the Mughals, turned India into the biggest importer of foreign bullion outside Europe.[25] Foreign supplies and pre-existing stocks were converted into money and integrated into the economy through the twin institutions of the mint and the market.[26]

Monetary metals were supplied to the mint in the form of Indian and foreign coins and uncoined metals shaped in bars or craft objects.[27] At the mint, the price of the incoming metal was calculated from its weight (*wazn*) and fineness (*aiyar*).[28] The weight of the fine metal (gross weight less alloy) determined the total number of coins of uniform weight and fineness which could be struck from it. The total output determined the mint charge and the amount delivered to the supplier (mint price or *dadani*). The supplier was issued a schedule of payment reckoned in a unit (*mal*) representing 1000 coins.[29]

Table 1

Mint Charges for Coining Money: AD 1578–92
(Percentage of total output)

	Muhrs	*Rupees*	*Dams*
Wages	0.24	0. 25	2. 85
Raw materials	0.50	0. 03	1. 31
Seigniorage	5.91	5. 00	5. 00
Total charges	6.65	5. 28	9. 16

Source: Based on *Ain-i Akbari*, I, pp. 31–33; tr. pp. 38–39.

The mint price of the monetary metal (number of coins delivered by the mint out of a given quantity of fine metal) determined the level of supply to the mint and the output which could be raised with the price of the metal. This mint price could be raised in a number of ways (e.g., short run debasement) but reduction in the mint charge was the most effective.[30] The mint charge comprised brassage-wages (*ajura-i-karigaran*) and the cost of raw material and seigniorage (tax on coining money). The biggest component of the mint charge was seigniorage (*wajib-i-sarkar* or *wajibi*) which the state levied to pay the salaries of the mint officials, raise revenue and safeguard the monetary usage of the legal tender.[31] The charges for coining the gold *muhr* (1.02 *tola*/188 grains troy), silver *rupiya* or rupee (0.95 *tola*/177.94 grains troy) and copper *dam* (1.75 *tola*/323 grains troy) are given in Table 1. The mint charge and the commission of the *sarraf* determined the market value of the coin in terms of the metal itself ('premium') as well as its exchange rate against other coins.

Standards of Fineness and Methods of Assay

The first step in the minting operation was to calculate the purity of precious metals (assay). In India, at least from the fourteenth century, gold was assayed on a standard of 12 degrees of fineness (*ban*). On this standard, the gold of highest purity was called *barabani* and was understood to be free of alloy. The calculations involved in the assay were based on an easy correspondence between the units of weight and the degree of fineness, like the carat system.[32] In the late sixteenth century, it was discovered that the *barabani* gold was not absolutely pure and contained alloy which could be refined away. As a result, the old standard was downgraded by 1.5 degrees so that the *barabani* gold reached only 10.5 *ban* on the new scale.[33]

Silver was assayed on a scale of 20 degrees (*bist-biswa*) of which, it seems, only the top 13 were used for testing in the market and the mint.[34] Although copper was never graded for its fineness in the same way as gold and silver, there was a notion of 'best copper' (*mis-i jaiyid*) for metal which had lost impurities and alloys during smelting.

Qualitative Assaying by Touchstone

The assayers in the mint or the bullion market sometimes gauged the fineness of precious metal coins simply by the colour of its surface and lustre. To the great annoyance of the present-day coin collectors, scrupulous money-changers pierced holes to look inside as well.[35] A combination of ocular and metallurgical means was the method of beating the silver hot and cooling it in water to discover whichever alloy prevailed: blackness denoted lead (*surb*), redness copper (*mis*), greyish whiteness zinc (*ruh-i tutia*), and a large proportion of whiteness denoted silver (*nuqra*).[36]

The more reliable method for qualitative assaying was called *banwari*. This was based on testing gold and silver on the touchstone (Hindi *kasauti*; Pers. *sang-i mihakk*).[37] In this time-honoured technique, lines were drawn with the metal on the touchstone and the colour of the mark (Skt. *varna*; Hindi *ban*) was judged against the mark of a needle of known composition.[38] Abul Fazl describes *banwari* as well as the preparation of touch-needles to assay gold from the highest to the lowest mint standard.

Although in this country, quick-sighted *sarrafs* (*sayrafian-i didawar*) can recognise the degree of fineness (*paya-i aiyar*) from experience by the colour and brightness (of gold), for the satisfaction of one and all this excellent method came into use. At the end of a few needles (*qalam*), made of copper or a similar metal, small pieces of gold (*tila*) are affixed with degrees of fineness inscribed on them. When a fresh piece of gold is brought for assay, a few lines are drawn with it and with the needles on the touchstone (*sang-i mihakk*). Whichever (mark of the touch-needle) comes closest is considered to be the degree (of fineness of the gold). However, (all) lines should be drawn with the same strength and in the same manner so as not to create a false impression.[39]

Table 2

Mint Standard of the Fineness of Gold: c. 1595

Degree of fineness (ban)	*Pure gold per unit of a given weight*	*Alloy per unit of a given weight*	*Percentage of pure gold*	*Percentage of alloy*
10.50	64.00	0.00	100.00	0.00
10.25	63.00	1.00	98.44	1.56
10.00	62.00	2.00	96.88	3.12
9.75	61.00	3.00	95.31	4.69
9.50	60.00	4.00	93.75	6.25
9.25	59.00	5.00	92.19	7.81
9.00	58.00	6.00	90.63	9.37
8.75	57.00	7.00	89.06	10.94
8.50	56.00	8.00	87.50	12.50
8.25	55.00	9.00	85.94	14.06
8.00	54.00	10.00	84.38	15.63
7.75	53.00	11.00	82.81	17.19
7.50	52.00	12.00	81.25	18.75
7.25	51.00	13.00	79.69	20.31
7.00	50.00	14.00	78.13	21.87
6.75	49.00	15.00	76.56	23.44
6.50	48.00	16.00	75.00	25.00
6.25	45.00	19.00	70.31	29.69
6.00	42.00	22.00	65.63	34.37

Note: Pure gold here means gold of 10.5 *ban*.

The various grades of gold to be affixed to the tips of the needles were obtained by combining 'pure' gold of 10.5 *ban* with alloyed gold of 6.5 *ban* in fixed proportions. First, a tri-metallic composition was prepared by mixing six *mashas* of 'pure' gold with one *masha* each of pure silver and copper (75% of gold and 25% of the alloy of silver and copper). From the composition, one *masha* was cut and divided into sixteen equal parts of half *ratti* (1 *masha* = 8 *rattis*). The combination of 'pure' gold of specific weight (from 7.5 to 0.5 *ratti*) with one or more parts of the composition in units of one *masha* yielded the gold of specific fineness. For example, 7.5 *rattis* of 'pure' gold combined with one part of the composition (0.5 *ratti*) to produce the gold of 10.25 *ban*. In this way, the addition of each part of the composition reduced the fineness of gold by a quarter *ban*. Thus, 7 *rattis* of 'pure' gold combined with two parts of the composition (1 *ratti*) to produce gold of 10 *ban*. The touch-needles below 6.5 *ban* were prepared using a different method. Six *ban* was the lowest limit of *banwari*, and gold falling below this standard was not accepted by the mint (Table 2).[40]

The touch-needles described by Abul Fazl assayed gold only up to 10.5 *ban* of the new standard and it is not clear how the test for higher fineness was performed. Sarma has suggested that additional touch-needles could not be made because the gold available in the market was of the old stock.[41] This would mean that qualitative assaying was restricted to 10.5 *ban* even in the mint and a different method was used to test gold of higher fineness.[42]

Qualitative assaying was easy, quick and inexpensive and was used as preliminary and confirmatory tests for smaller quantities of bullion, coins and objects of known fineness. For greater precision in assaying a wider variety of precious metals, the mint had to supplement it with an altogether different method.

Quantitative Assaying by Refining

Quantitative assaying was the method of testing fineness by calculating the loss of weight suffered by the metal during refining with the understanding that the quantity lost constituted the alloy burnt in the fire while the remainder was the pure metal.[43] Thus, if the gross weight of the metal was 100 and the net weight after refining was 90, then the alloy was 10 per cent and the pure metal was 90 per cent. The fineness could be stated in degrees of the scale where each *ban* represented a fixed weight of pure metal.

The test required the combined skills of an assayer (*sarraf*), weigher (*wazn kash*) and scribe.[44] The metal brought to the mint was melted, thoroughly mixed and cast into bars of uniform consistency.[45] Samples were cut, weighed, refined using the standard techniques and weighed again.[46] The difference between the gross and net weights of the samples indicated the fineness of the metal. The figure had to be calculated with extreme precision since it determined the worth of the entire quantity and a small error could make a big difference to the mint and the supplier. As mentioned earlier, the figure yielded by assaying provided the basis for calculating the mint price and the mint charge. It also served another purpose.

The relationship between weight loss and the proportion of alloy guided the process of refining insofar as the refiners based their calculations on the figure supplied to them by the assayers. The refiners had to know in advance, the quantity of impurity to be burnt in order to bring the metal to the required degree of fineness. For instance, if 100 *tolas* of gold were brought to the mint which touched 9.5 *ban* and burnt 6 *rattis* in one *tola* sample, then the refiners would know that 6.25 *tolas* had to be burnt for the entire gold to reach 10.5 *ban*. Thus, the work of the departments of assaying and refining complemented each other. The limitation of this relationship was revealed when an irregularity in the mints was discovered in the late sixteenth century which had implications for the monetary authorities as well as money users.

When gold was refined, a small amount of it was lost in the dross (*niyara*) which was later recovered, if the mint had the right technology. The mint could do one of two things with the dross. It could sell the dross to make money. This was the practice in the Orissa mint, of which we have the record, where the dross of gold was retained by the refiners on payment of a fixed monthly fee to the mint (*peshkash-i niyara*).[47] Alternatively, the mint could recover the gold, credit the supplier with the amount and levy an additional charge to cover the cost. This appears to be the standard practice in Mughal mints after the fraud incident. Until then, it appears that the workmen in the department of refining retained the dross and defrauded both the mint and the supplier. When the fraud was detected, those who were dishonest (*khiyanat peshgan*) were disciplined and the practice was put to an end.[48]

The discovery of the fraud led to the realisation that if the dross contained not only impurities but also some gold, then the sample assayed by refining must have retained the same quantity of alloy as the gold lost in the dross in order that the gross and net weights balanced up. In other words, after the estimated loss of weight had occurred, the pure gold in the above sample weighing 100 was 90 (and not 93.75) whereas the impurity in the dross was 2.50 (and not

6.25). Refining techniques must be either improved to bring the fineness of the gold to the higher standard. It is not surprising that Abul Fazl waxes lyrical on the purity of gold achieved in the Mughal mints and mentions with pride the feat accomplished under his patron.[49] The revision of the gold standard can also be seen in this context.

The Technology of Refining Gold and Silver

Tavernier, the peripatetic French jeweller who visited India in the seventeenth century, mentioned that precious metals were refined to the highest standard in the Mughal Empire before they were coined into imperial money.[50] This was a common belief amongst almost all the commentators on Mughal coinage in the seventeenth century, borne out by the high fineness of the coins themselves (Table 5), and by references to the process of refining of gold and silver in every mint record or manual. The following description is based mainly on the chapters of Abul Fazl's book on the method of 'refining alloyed gold and silver' and the various 'separative processes'.[51]

After the net worth of the monetary metal was evaluated by the mint officials, the metal was sent for refining in the department of smelting (*gudazkhana*). The furnace used for refining, with the notable exception of a clay vessel, was a hole dug in the earth, the size of which depended on the quantity of the metal. Charcoal (*ingisht*) was used as fuel for a strong fire and dried cakes of cow dung (Hindi *upla*; Persian. *pachak i dashti*) for a gentle fire.[52] Draught was blown with skin-and-wood bellows (Hindi *dhunkni*; Pers. *dama*) operated manually by the smith or his assistant.[53] Metals were melted and cast in crucibles (*buta*) and clay moulds (*gilin takhta*). Lead was the most important ingredient for refining and constituted 70 per cent of the total cost (Tables 3 and 4).

Gold coins and ornaments containing alloys of silver and copper were refined by cementation. The unrefined gold was first cast into bars by the smelter-cum-refiner (*gudazgar-i kham*) and beaten into thin square plates by the plate maker (*warq kash*).[54] The plates were measured by the master assayer in a copper mould and stamped to verify their identity and to guard against 'fraudulence'.[55] The stamped gold plates were washed with clean water and interleaved with the cementation compound (*daru*) of saltpetre (*shora namak*) and dust of raw brick (*khisht-i kham*). The composition was covered with *upla* and set on fire, which should not to be too strong to melt the gold. The fuel burnt slowly and gently (*ahista ahista*) until it turned into ashes. The saltpetre burnt the copper and separated the gold from the silver while the earthy matter absorbed the admixtures. When three successive fires had been applied, the plates were washed and cleaned.

Cementing could be done more than once if gold was to be refined to higher degrees. In that case, the plates were burnt again with a fresh compound. After eighteen fires had been applied to six compositions, the refiners submitted the purified plates for inspection. The master assayer broke them one after another. A soft and gentle sound indicated purity, while a harsh sound subjected the gold to more fires. Finally, a small quantity (one *masha*) of gold was sliced off each refined plate and forged into a sheet which was assayed on the touchstone. If need be, quantitative assaying was also performed. Twenty plates each of refined and pure gold of uniform weight and shape were burnt with the cementation compound and weighed. If the test plates were found to be equal in weight with the check plates, they were considered pure. If the gold was still not sufficiently pure, it passed through one or two more fires before it was approved.[56]

Residual gold absorbed by the scoria (Hindi *saloni*; Persian *khak-i-khalas*) during refining was recovered in two stages. The bulk of the gold was recovered in the first stage by amalgamation in which small quantities of *saloni* were washed and rubbed with mercury (*simab*). The mercury, due to its 'predilective affinity (*jazb-i muhabbat*)', was able to suck up the gold settled at the bottom of the container to form an amalgam. The amalgam was heated in a retort (*shisha*) to evaporate the mercury and obtain the pure gold. The mints employed scoria-sifters (*niyariya*) for the recovery of gold from the dross.[57]

A more elaborate process was used in the second stage to retrieve whatever gold, silver and copper that still remained in the washed-away *saloni*, called *kukra*. In the first phase of this stage, the metals were separated from the scoria with the help of lead. For this, the *kukra* was pounded and mixed with specially prepared lead soaked in ashes, called *punhar*.[58] The mixture was worked into balls with a paste made from a mild acid (*rasi*) and cow dung cakes.[59] A clay furnace, shaped like a jar but open and narrow at both ends and wide in the middle, was filled with charcoal almost to the brim and placed over a pit. The charcoal was lit and a strong fire was blown with a pair of bellows. The balls were broken into pieces and dropped into the furnace. Molten gold, silver, copper and lead fell through the bottom of the vessel into the pit below. The bulk of the lead was, however, retained by the ashes in the furnace and later separated and reused. The ashes were not thrown away since silver particles could still be recovered from them. The lead in the pit was burnt and separated from the metals by using the same technique of binding it with the ashes as was used to make *punhar*. The final product was a tri-metallic mass of gold, silver and copper (*bagrawati*).

In the second phase, copper was separated from gold and silver by fusion of the metals themselves. The *bagrawati* was heated twice in a dish (*rikebi*) shaped from the ashes of the *babul*-wood (*accacia Arabica*; Persian *mughailan*) with the addition of copper and a larger quantity of lead (in a ratio of 1:25). The copper and the lead united with the ashes, leaving the gold and silver together. The ashes left behind were retained to recover the lead and copper by using a different technique.

Table 3

The Cost of Refining 100 Tolas of Gold: AD 1578–92

Workmen and items	*Wages and costs of raw materials (rupees)*	*Percentage of the total cost*
Refiner	0.06	1.08
Plate maker	1.05	18.89
Scoria-sifter	0.50	8.99
Cow dung cakes	0.51	12.72
Saltpetre and brick dust	0.09	2.29
Water	0.02	0.51
Mercury	0.21	5.34
Charcoals	0.40	10.18
Lead	2.72	68.96
Total cost	5.56	100.00

Source: *Ain-i Akbari*, I, pp. 16, 19–20, 31–32.

In the third phase, the precious metals were separated from each other. The mixture of gold and silver was broken into small particles and melted six times in a crucible: thrice with copper and thrice with sulphur (Hindi *chhachhiya*; Persian *gugird*). In the intervals, the heated particles were dropped in cold water covered with *khas* straws (a sweet-scented grass) to solidify. The water was constantly stirred with a wooden stick to prevent the particles from reuniting into a mass. After the intermittent heating and cooling process, the gold was refined away from silver. The gold (*pinjar*) obtained from this process was of a low fineness (between 4 and 6.5 *ban*) and could be refined up to 9 *ban* either by the standard *saloni* process or its variant called *aloni*. The residue (*kharal*) was retained to recover particles of silver through cupellation.

Mint silver brought with the alloys of lead, zinc (*jast*) and copper was refined through cupellation with the addition of lead (*sabbaki*). The refiner (Hindi *shoddha*; Pers. *sabbak*) put the mixture of silver and lead in a furnace in a dish, lined with the ashes of *upla* and *babul* wood. The mixture was covered with charcoal and a strong fire was blown with bellows till the mixture was melted. The entire process was generally repeated four times.[60] Approaching the final stage of refining, molten silver radiated brightness and began to harden first at the edges and then in the middle. When a few drops of water were sprinkled, excrescences (*shakh-ha*) appeared on the surface and the silver was shaped like a disc (*qurs*).[61] This was the sign that the silver had reached the required degree of fineness.[62] Refined silver was heated and hammered by the beater (Hindi *chandipit*; Pers. *qurs kob*) to get rid of the remaining lead.[63] The discs of refined silver were marked by the master assayer with an official stamp (*sikka-i adl*) to guard against alterations.

The ashes (*kharal*) collected after the process of refining was over contained the dross (*murdar sang*) of silver, lead and copper. The *kharal* was pounded with mortar and pestle and kneaded with borax (Hindi *sohaga*; Pers. *tangar*) and natural sodium carbonate (*ashkhar-i kofta*). The mixture was subject to the same treatment as the *kukra* in the first phase of the separative process. While the copper was burnt away, the silver and lead were collected in the pit. The silver was purified from the lead by *sabbaki*. The silver scoria-sifter (*paniwar*) was not a regular employee of the mint and charged the suppliers directly for raw materials and labour. For plying his trade inside the mint the *paniwar* contributed a fixed monthly sum to the exchequer.[64]

Table 4

The Cost of Refining 100 Tolas of Silver: 1578–92

Workmen and items	*Wages and costs of raw materials (rupees)*	*Percentage of the total cost*
Refiner	0.14	18.18
Beater	0.01	1.30
Scoria-sifter	0.06	7.79
Lead	0.54	70.13
Charcoals	0.02	2.60
Total	0.77	100.00

Note: Scoria-sifter's wages are not separately listed but appear to have been clubbed with those of the refiner in the account of the cost of minting. The wages of the refiner alone (a quarter lower) are given in the account of the minting operation.

Source: Ain-i Akbari, I, pp. 18–19, 32–33.

We have figures for the late sixteenth century of the values of gold and silver recovered from the scoria and the cost of refining. The loss of weight in refining 100 *tolas* of gold of 8.25 *ban* was 17.5 *tolas*. The quantities of gold (of the standard of the *muhr*) and silver recovered from the ashes were 2.24 *tolas* and 9.08 *tolas* respectively.[65] The value of the two together was 27 rupees whereas the total cost of refining was 5.56 rupees.[66] The refining technology available at the mint was therefore cost effective and contributed towards raising the mint price of gold. The cost of refining 100 *tolas* of silver was less than a rupee but it was double the amount recovered from the ashes (0.35 rupees). However, the workshop in the mint was used for commercial purposes, and gold and silver were refined in the mint as well as recovered from *saloni* and *kharal* bought from the goldsmiths of the city.[67]

The Fineness of Mughal Coins

After the gold and silver had been refined, they were tested by the assayer (Hindi *chauksi*; Pers. *chashnigir*) for approval for minting. For quantitative assaying the method used was once again trial by fire. Samples (*chashni*) of the refined gold were burnt with the cementation compound and weighed. If no loss of weight occurred, the gold was supposed to have reached the required standard. The results were verified by testing the samples on the touchstone before the master assayer. Similarly, a sample of the refined silver was cut weighing one *tola* and heated with the same quantity of lead in a bone crucible (*buta-i ustukhani*) keeping the draught out. When the lead was separated, the silver was melted in another crucible. If the sample lost three rice grains in a *tola*, which it did during casting the silver was considered to be sufficiently pure and was tried on the touchstone.[68]

Before the approved gold and silver were passed on for actual coining, their fineness was adjusted by adding a small quantity of alloy.[69] Neither of these metals was suitable for direct employment in making coins owing to softness and lack of resistance to wear and had to be alloyed to produce hardness and workability. Abul Fazl himself gives a clue to this when he mentions a small deduction (0.46%) made for the *khalisa* from the ready-to-be-minted silver that remained unaccounted for in the balance sheet of the mint.[70] If the deduction was made to add alloy for industrial usage, then the rupee was struck to very high fineness (99.54%) in the sixteenth century. Numismatic and textual evidence suggests that the alloy content of the rupee marginally increased, and its fineness declined, in the next century.

In a 1996 laboratory test, the British Museum analysed 10 specimens of Mughal rupees emanating from four mints (Agra, Ahmadabad, Lahore and Surat). The selection of mints was intended to give a broad coverage of the Empire from AD 1569–70 to 1660–61 with one major mint, Agra, represented by five coins with a range of dates. All the coins were analysed using two methods: energy dispersive X-ray analysis (EDX) in a scanning electron microscope (SEM) and energy dispersive X-ray fluorescence (XRF). Both methods require the exposure of a representative area of uncorroded metal for analysis. This was achieved by abrading and polishing a small area on the edge of each coin to remove surface layers altered by corrosion and through coin manufacture. The results (Table 5) were obtained by taking the mean of the two methods for silver and copper but using the gold and lead (generally not detectable by SEM-EDX) figures produced by XRF.[71]

The laboratory results show that all rupee coins have very similar silver content (fineness) in the range of 97.9–99.2 per cent with an average of 98.8 per cent. The results also indicate a very high degree of fineness and uniformity of official standard across the regions of the Empire. Traces of three alloys were found in the silver of the rupee, viz., lead, gold and copper. The lead content of the coins is derived from the ore type (silver was extracted from argentiferous lead ores) and the lead added during the refining process (cupellation). There seems to be no systematic difference in the lead content of the coins from the various mints, although coins from the major mints, Agra and Lahore, tend to have somewhat lower and less variable lead content which might indicate a more controlled process of refining.

The cupellation refining removes most of the baser metal from silver but does not extract gold; hence the small amounts of gold which are commonly associated with silver are carried through the process intact. The gold content ranges from undetectable (0.07%) to 0.73 per cent and there is some indication of differences among the mints. Thus, coins from the major mints of Agra and Lahore have comparatively high gold content (0.41–0.73%) whereas those from the other mints contain, at most, 0.2 per cent of gold. This suggests that the stocks of silver used at Agra and Lahore were different from the other mints, although it is not possible to indicate sources from these data.

The fineness of coins had important political and economic implications insofar as a strong currency had a greater degree of acceptance and mobility. The reputation commanded by the Mughal rupee on account of its unvarying weight and fineness expanded its

Table 5

The Fineness of Mughal Silver Coins (Rupees)

Ruler	*Mint*	*Date (AH)*	*Date (AD)*	*Silver (%)*	*Copper (%)*	*Gold (%)*	*Lead (%)*
Akbar	Agra	977	1569–70	99.10	0.30	0.54	0.06
Akbar	Lahore	989	1581	99.20	0.20	0.53	0.11
Akbar	Ahmadabad	1001	1592–93	98.70	0.80	0.08	0.08
Jahangir	Agra	1026	1617	98.70	0.90	0.46	0.21
Jahangir	Agra	1033	1623–24	97.90	1.00	0.07	0.93
Jahangir	Surat	1035	1625–26	98.90	0.50	0.73	0.24
Shahjahan	Agra	1038	1628–29	99.20	0.30	0.41	0.18
Shahjahan	Surat	1067	1656–57	98.60	1.10	0.20	0.20
Aurangzeb	Surat	1071	1660–61	98.60	0.80	0.10	0.47
Aurangzeb	Agra	1071	1660–61	98.80	0.80	0.20	0.11

Source: M. R. Cowell, *MS. Report of Energy Dispersive X-ray (EDX) and Energy Dispersive X-ray Fluorescence (XRF) Analyses of Mughal Silver Specimens*, Department of Coins and Medals, British Museum, 1996.

area of circulation beyond localised points of production. The specimens of Akbar's rupee recovered from the hoards discovered in the United Provinces (Uttar Pradesh) are illustrative of the inter-regional mobility of the silver coin. Of the 1,967 coins, 81 per cent came from mints situated outside the map area of UP (excluding Delhi which lay at the western border). One reason for this was the gravitational pull of the area which housed the imperial capital (Agra) and the central exchequer. But the other, and perhaps more important, factor was the presence of markets which supplied goods for long-distance trade. The rupee by moving across regions and localities ensured that networks of exchange operated without frontiers throughout the Empire. When the fineness of the rupee declined in the eighteenth century, its circulation was localised.[72]

Fabrication of Blanks

The penultimate stage in the minting operation was the preparation of plain metal pieces shaped like coins (blanks). Sheets of standard gold, silver and copper were cast into strips (*shosha*) suitable for cutting blanks. The strips were cast in a clay mould either by the caster (*gudazgar-i pukhta*) or the refiner (*sabbak*). In order to prevent the metal from sticking to the mould, a layer of grease (*charb*) was applied in the case of precious metals and ashes in the case of copper.[73] As mentioned earlier, the silver lost a fraction of its weight during casting (*haqq un nar*) and the loss was debited to the account of the supplier. It amounted to half a per cent in the late sixteenth century but a little less in the later two centuries.[74]

Malcolm's description of this workshop suggests that the technique used for casting strips was the same but the loss of weight now no longer occurred:

> The purified mass is afterwards taken to the melter, who putting one thousand rupees weight at a time in a large crucible on an iron ring, capable of being raised by attached chains, melts it and runs it into several small flat moulds about six inches long, and half an inch broad, forming it thus into convenient pieces for cutting into the necessary dimensions.[75]

The *zarrabs* cut the blanks (Hindi *chakli*; Pers. *mutallas*) from the three metal strips.[76] Their task was delicate and demanded precision since the blanks were to be cut to an equal size an ' weight (*kamil ul wazn*) for each denomination.[77] Abul Fazl noted that, unlike Iran and Central Asia, the Indian *zarrabs* prepared blanks without the help of the scale (*mizan*).[78] Cutting square blanks may not have been that difficult and Abul Fazl was writing at a time (AD 1578–92) when only square coins were minted. In nineteenth century Central Indian mints, anvils, shears and scales were used by a team of three craftsmen to fashion spherical blanks from silver bars:

> The bars of silver are then delivered to the silversmiths, each of whom has a small raised fireplace and anvil in front close to him. On one side sits another with scale and shears, for supplying him with square pieces of the metal with nearly the proper weight. On the other side is a person whose business is to adjust the weight more accurately after it has been formed into its shape. The silversmith receives back the small lumps; heats them red hot, and taking them up with a pair of small forks, gives them two or three smart blows on the angular points, then strikes the piece flat, and gives it afterwards one or two rapid turns on its edge, accompanied by gentle strokes of hammer; and it thus receives its rudely round form ready for the die.[79]

Coin specimens preserved in museum and private collections betray – advancing a margin for wear and tear, and tolerance (permissible deviation from the mint standard) – a high degree of uniformity in weight, size and shape within a single series issued by the mint. In the late sixteenth century, wages paid to the *zarrabs* were the highest of all mint workmen: 79 per cent of the total wages for coining rupees out of fine silver and 62 per cent for coining copper *dams*.[80]

Die-Engraving

In medieval Indian mints, as indeed in mints the world over, dies were used for stamping coins. The dies were usually of iron or steel (*fulad*), and comprised two parts: the upper part, called punch or trussel (Pers. *musala*), and the lower part, called pile (*dasta*!).[81] The lower face of the punch and the upper face of the pile were engraved with inscriptions (words and figures) sanctioned by the monetary authority.[82] The inscriptions which were more permanent in nature, such as the *kalima*, names of the caliphs and emperors, benedictory formulas and regal protocols, were usually engraved on the pile and appeared on that side of the coin which the numismatists call obverse. Those which were modified periodically, such as the *hijri* and regnal years of mintage, were engraved on the punch and appeared on the reverse.[83] Apparently, this was done because the punch, due to the blows it received from the hammer, wore out faster than the pile and needed to be replaced more frequently.

The die-engraver (*muhrkun*), for his artistic skills and knowledge of calligraphy, appears a less shadowy figure in our sources. He cut the inscriptions on the face of the dies which remained in contact with the blanks. Maulana Maqsud Hirawi and Maulana Ali Ahmad Dehalwi were supposedly two eminent engravers and unrivalled practitioners of their craft in the late sixteenth century. Engravers were ranked officials and had assistants working for them in the mint.[84] Unfortunately, not many Mughal dies have survived, and there is practically nothing in the contemporary literature on the use and distribution of dies. Dies in the medieval mints were well-guarded treasures to which nobody was allowed access for fear of imitation and counterfeiting.[85] In the Mughal Empire, cases of counterfeiting were understandably rare given the high fineness of coins and the widespread custom of assaying.[86] Counterfeiting was also a breach of royal prerogative and it is perhaps for this reason that the Mughals did not allow die imitation. The dies may have been manufactured and cut locally, but at the time of the extension of the imperial currency system to the conquered territories, dies were sent from the capital rather than allowed to be imitated.[87]

Die-Striking

The final operation in the process of minting was stamping the blanks. The blanched blank was placed between the two dies, the pile fixed to the ground (or a block of wood) and the punch suspended over the blank. The punch was struck with a hammer (*ghan*) by the coiner's assistant (Hindi *ghatwara*; Pers. *patakchi*) to stamp both sides simultaneously.[88] The job of the coiner (*sikkachi*) was to fix the dies, place the blank between them and remove it after striking, examine the impressions formed on the obverse and the reverse, and prevent slipping, over-striking and cracking of the edges of the coin. In addition, he was required to supervise the repair of the hammer and the dies which belonged to the mint.[89] In most Mughal specimens, especially those struck out of precious metals, the entire legend can be seen on the face of the coin (flan). The legends inscribed on the margins of the coin could be lost (off flan) either due to a mismatch between the size of the die-surface and the blank or sloppy striking. The operation demanded more skill and responsibility than the hammerer's job, which was to strike the blow with measured strength at the instruction of the coiner. The hammerer received a share of the wages paid to the chief coiner.[90]

Since Mughal coins were mostly round in shape, the face of the dies too was round which may have made their alignment difficult (it would be easier if the dies were square-faced). From an examination of museum specimens, where the inscriptions on the obverse and reverse of the coins are not always fixed in relation to each other (hold the obverse straight and flip the coin to find that the inscription on the reverse is not perpendicular), it appears that the Mughal dies were not pegged and no particular care was taken for die alignment in stamping the blanks. Given the scale and magnitude of the Mughal mint production, the hammerer and his assistants may have been required to work at a pace which obviated the necessity of die alignment.

Conclusion

Coin production in Mughal mints was organised, systematic and uniform although there might have been regional variations in the execution of skills. The raw materials passed through the hands of at least 10 teams of mint workmen in a long departmental chain before the finished product was approved for use. Quality control was rigorous and mint officials were held responsible for maintaining high standards. The technique of stamping coins with dies and hammer, known as die-striking or hammer striking, was widely prevalent in the medieval world and remained in use in India until it was replaced by milling in the nineteenth century. The technology was manual and based on the use of simple tools and devices. There was no use of animal power even for driving bellows to ignite strong fires. As a result, the furnaces and crucibles were not capable of melting large quantities of metal. The technology was labour-and-skill intensive and relied heavily on the long-standing traditions of smelting, refining and coining prevalent in India.[91] Major changes in the recruitment of labour force and the mint buildings were not frequent and, except for improvements in refining precious metals, no technological breakthrough took place to bring production in line with the supplies. The technology was put to the test during high seasons when the slow working of the mint invited complaints from the merchants. The positive side of this was the growth of the money market and the business of the money-changers (*sarrafs*), who turned to offering cash and credit facilities to merchants for investments at the coasts as well as the distant hinterland markets.

Notes

1. Najaf Haider, 'The Quantity Theory and Mughal Monetary History', *Journal of Medieval History* 2, ii, 1999, pp. 328–29.
2. Najaf Haider, 'Precious Metal Flows and Currency Circulation in the Mughal Empire', *Journal of the Economic and Social History of the Orient*. Special issue on 'Money in the Orient', 39 (1996), pp. 326–42; Om Prakash, 'Foreign Merchants and Indian Mints in the Seventeenth and Early Eighteenth Century', *The Imperial Monetary System of Mughal India*, John F. Richards (ed.), Delhi: Oxford University Press, 1987, pp. 173–74, 189.
3. C.R. Singhal, *Mint Towns of the Mughal Emperors of India*, Bombay, India: Numismatic Society of India, 1953; John S. Deyell, 'The Development of Akbar's Currency System and Monetary Integration of the Conquered Kingdoms', *The Imperial Monetary System of Mughal India*. op. cit., pp. 58–59.
4. Abul Fazl, *Ain-i Akbari*, 2 Vols., H. Blochmann (ed.), (Bib. Indica, 1872–77), I, pp. 12–23; tr. H. Blochmann, D.C. Phillot (ed.), Calcutta, 1927, pp. 16–27. The Mughal chancery used the term *dar uz zarb* for the mint while it was commonly called *taksal* in the Indian languages. See, *English Factories in India* (1618–69), 13 Vols. (each volume titled by the year/s it covers), William Foster (ed.), Oxford, 1906–27, 1618–21, p. 293; *John Marshall in India. Notes and Observations in Bengal 1668–1672*, Shafaat Ahmad Khan (ed.), London, 1927, p. 117. In a comprehensive *Latin-Hindustani-French Dictionary*, (Bibliotheque Nationale, Paris, MS. Indien 839, ff. 73–4) compiled in India in the late seventeenth century, the term 'tangsal' is rendered as 'Maison de la Monnoye' (mint).
5. *Ain-i Akbari*, I, pp. 31–33 (*Ain-i sud-i bazargan-i tila wa nuqra*); tr. pp. 38–39.
6. *Dastur ul Amal-i Alamgiri*, Oriental and India Office Collection (OIOC), British Library, London, MS. Add., 6599, ff. 53b–58a. The manual itself was written in 1659, but the record belongs to the year 1657.

7. Hidayatullah Bihari, *Hidayat ul Qawaid*, Maulana Azad Library, Aligarh Muslim University, MS Abdus Salam Collection 379/149, ff. 37a–39a; tr. William, *Proceedings of Asiatic Society of Bengal* (May 1898), text, pp. 149–50; tr. pp. 150–52. The Abdus Salam MS contains additional information on coin production omitted in the MSS used by Irvine; *Kaghzat-i Mutafarriqa*, OIOC, MS Add., 6586, ff. 57a–58b.
8. The revenue of the mint was farmed in the seventeenth century but not the minting operation and there is sufficient evidence to suggest that the revenue farmer had no control over the process of coin production. *English Factories, 1642–45*, pp. 23–25.
9. *Ain-i Akbari*, I, p. 15; tr. p. 20; OIOC, *Factory Records, Surat*, 5, f. 180a; OIOC, *Original Correspondence 6436*, f. 135b.
10. *Dastur ul Amal*, OIOC, MSS 1779, 1842, facsimile reproduction and tr. John F. Richards, *Document Forms for Official Orders of Appointment in the Mughal Empire* (Cambridge, 1986), ff. 223a–b; *Hidayat ul Qawaid*, ff. 37a–39a. For instructions which mint officials received from the emperor see *The Letter Book of Isaac Lawrence* (1677–79), OIOC, MS. Eur. E 387/A, f. 55b; *Factory Records, Surat*, 5, ff. 151a–b; *Factory Records, Dacca*, 1, ff. 182–3.
11. *Ain-i Akbari*, I, p. 15; tr. p. 20; *Hidayat ul Qawaid*, ff. 37a–39a; Ali Muhammad Khan, *Mirat-i Ahmadi*, 2 Vols. and *Supplement*, Syed Nawab Ali (ed.), Baroda, 1927–30, *Supplement*, p. 183.
12. *Dastur ul Amal-i Alamgiri*, ff. 53b–58a.
13. For references to the mint as a manufactory (*karkhana*) see, *Kaghzat-i Mutafarriqa*, ff. 57a–58b; *Zawabit-i Alamgiri*, OIOC, MS. Or. 1641, f. 134a. For a wide-ranging discussion of the mint as a workshop see Frank Perlin, 'The Parts of the Machine: Division of Labour in European and South-Asian Coin Manufacture before Mechanisation', *The Invisible City: Monetary, Administrative and Popular Infrastructures in Asia and Europe, 1500–1900*, Variorum, 1993, pp. 91–130.
14. Malikzada, *Nigarnama-i Munshi*, Nawal Kishore, 1882, pp. 144–53; *Document Forms*, f. 223b; tr. p. 47.
15. Irfan Habib, 'Potentialities of Capitalistic Development in the Economy of Mughal India', *Enquiry*, New Series, iii (1971); repr. in *Essays in Indian History: Towards a Marxist Perception*. Delhi, 1995, p. 222.
16. OIOC, *Home Miscellaneous*, H/357, ff. 429–30, 465–69.
17. *English Factories, 1642–45*, pp. 17–18; *Factory Records, Surat*, 94, f. 69a; 92, f. 1.
18. *English Factories, 1618–21*, pp. 7–8; *1646–50*, p. 101.
19. *Factory Records, Surat*, 94, f. 39a; *Home Miscellaneous*, H/357, ff. 429–430, 465–466,
20. *Factory Records, Surat*, 94, ff. 7b–8a, 154a–154b. The two lunar holidays were universally observed by merchants and craftsmen in the Mughal Empire. *Mirat-i Ahmadi*, I, p. 260 (1666 AD); *Zawabit-i Alamgiri*, f. 137a (c.

1691 AD); *Dastur ul Amal*, Bodleian Library, Oxford, MS Fraser 86, f. 38a (c. 1696 AD).

21. *Factory Records, Surat*, 94, ff. 7b–8a, 154a–154b.
22. See W.H. Moreland, *From Akbar to Aurangzeb: A Study in Indian Economic History*, London, 1923, p. 177 for an expansion in the mint output at Surat in response to merchant's demands. In April 1694, the Surat mint was expected to send additional workmen to Delhi to cope with the increased scale of production there. *Akhbarat-i Darbar-i Mualla*, Royal Asiatic Society, London, MS 2361 (24 *shaban*, 37th regnal year of Aurangzeb).
23. Najaf Haider, 'The Monetary Integration of India under the Mughal Empire', India, *Studies in the History of an Idea*, Irfan Habib (ed.), New Delhi: Oxford University Press, 2005, pp. 132–34; *English Factories, 1624–29*, p. 241; note 2 above.
24. There was practically no domestic extraction of gold or silver. Even though copper was extracted from the mines of Central India, its recurring economic demand outstripped supplies.
25. In the last quarter of the sixteenth century, the annual supply to the Mughal Empire was 124 metric tons of silver equivalent. Haider, 'Precious Metal Flows', p. 323. The estimates offered by Shireen Moosvi are 184.6 metric tons of silver and 4.6 metric tons of gold. *The Economy of the Mughal Empire c. 1595: A Statistical Study*. Delhi, 1987, pp. 375–76. Ming China was the second biggest importer with an annual intake of 43 to 46 metric tons in the same period. Richard Von Glahn, *Fountain of Fortune: Money and Monetary Policy in China, 1000–1700*. Berkeley: University of California Press, 1996, pp. 133–40.
26. Haider, 'Precious Metal Flows', pp. 300–35.
27. *Ain-i Akbari*, I, pp. 31–32; tr. pp. 38–39; *Dastur ul Amal-i Alamgiri*, ff. 54a–b, 55b, 57a; *Kaghzat-i Mutafarriqa*, ff. 57a–58b; Jean-Baptiste Tavernier, *Travels in India*, 2 Vols., tr. V. Ball, 2nd edn. rev. W. Crooke, London, 1925, I, pp. 11–12.
28. Bullion, base metals and coins were weighed on a scale of which the highest unit was the *tola* (Skt. *tula*, balance). Each *tola* was divided into 12 *masha*, 96 *ratti* or *surkh* and 576 (later 720) rice grains (Hindi *chawal*; Pers. *birinj*). *Ain-i Akbari*, II, p. 60. While the units of this scale remained constant throughout the Mughal period, the absolute weight of the *tola* varied as a result of different weight standards adopted by mint authorities. The weight of the *tola* in the sixteenth century was 185.5 troy grains. S.H. Hodivala, *Historical Studies in Mughal Numismatics*, Calcutta: Numismatic Society of India, 1923, p. 234.
29. *Ain-i Akbari*, I, pp. 31–32; tr. pp. 38–39; *Dastur ul Amal-i Alamgiri*, ff. 54a–57b; *Kaghzat-i Mutafarriqa*, ff. 58a–58b; *Surat Documents*, nos. 4 and 5, reproduced in M.P. Singh, *Town, Market, Mint and Port in the Mughal Empire, 1556–1707: An Administrative Cum Economic Study*, New Delhi: Adam Publishers and Distribution, 1985, Appendix; Pieter van Dam,

Beschryvinge van de Oostindische Compagnie, 4 Vols., 7 pts, F.W. Stapel and C.W. Th. van Boetzelar (eds.), (Gravenhage, 1927–54), I, ii, p. 80; *Original Correspondence 6436*, ff. 135a–b; *6490*, ff. 295b–296a; *Factory Records, Surat*, 5, ff. 170a–b.

30. *Mirat-i Ahmadi*, I, p. 265; Haider, 'Precious Metal Flows', p. 335.
31. The Perso-Arabic term *wajibi* was used for state dues in general in the medieval Indo-Islamic world. It appears from seventeenth century documentation that such a term was used to denote seigniorage in India. *Dastur ul Amal-i Alamgiri*, ff. 54a, 55b, 56b, 57a. For its use in the specific sense of mint dues outside India see Uthman Ibn Ibrahim Al-Nabulusi, 'Scandal in the Egyptian Treasury: A Portion of the Luma Al-Qawanin of Uthman Ibn Ibrahim Al-Nabulusi', ed. and tr. Charles A. Owen, *Journal of Near Eastern Studies*, xiv (1955), f. 101b; tr. p. 76; Mirza Samia, *Tadhkirat ul Muluk: A Manual of Safavid Administration (circa 1137/1725)*, facsimile reproduction of OIOC, MS Or. 9496; tr. V. Minorsky, London, 1943, f. 37b.
32. Sreeramula Rajeswara Sarma, 'Varnamalika System of Determining the Fineness of Gold in Ancient and Medieval India', *Aruna-Bharati: Professor A.N. Jani Felicitation Volume*, Baroda, 1983, pp. 369–389; V.S. Agarwala, 'The Highest Purity of Gold in India', *Journal of the Numismatic Society of India*, xvi (1954), pp. 271–74.
33. *Ain-i Akbari*, I, pp. 13–14; tr. p. 18.
34. *Ain-i Akbari*, I, p. 18; tr. p. 23; Tavernier, *Travels in India*, I, pp. 28–29; *Dastur ul Amal-i Alamgiri*, f. 77a. The reduction in the volume of previously minted billon coins delivered to the mint may have restricted the assay of silver to the top degrees.
35. *Ain-i Akbari*, I, p. 18; tr. p. 23. '(A)s for the sherofs (*sarrafs*) piercing holes in the dollars (reales of eight), however unusual in Europe, it is accustomary in these parts; nor can it be esteemed unreasonable, since the false can by no other means be discovered, and that good silver is due unto them'. *English Factories, 1646–50*, pp. 6–7.
36. *Ain-i Akbari*, I, p. 18; tr. p. 23.
37. An early seventeenth century *English-Persian Dictionary* (Bodleian, MS Hatton Or. 32), defines the touchstone as *sang-i mihakk*, while *An Anonymous Eighteenth Century Persian-Urdu Vocabulary* (Bibliotheque Nationale, MS Indien 819, f. 23) defines (*sang-I*) *mihakk* as *kasauti*.
38. 'They (the *sarrafs*) have got an excellent method and very easy, so that one cannot mistake the value of the denomination of gold and silver. They use touch-needles, all in one string made of brass, at the end of which is a small piece of gold or silver from the lowest alloy to the highest alloy with the price (*le prix*) mark inscribed on it. Thus, during the assay of your coins on the touchstone (*pierre de touché*) and making them equal to the touch, you can look at the value marked on the blades'. Georges Roques, *La maniere de negotier dans Les Indes Orientalles dedice a mes Chers amis Et Confreres Les Engages de la Royalle Compagnye de France*, Bibliotheque

Nationale, MS Fonds Francais 14614, f. 247. The use of touch-needles to assay purity was quite widespread in medieval mints. See, Andrew S. Ehrenkreutz, 'Extracts from the Technical Manual on the Ayyubid Mint in Cairo', *Bulletin of the School of Oriental and African Studies*, xv (1953), pp. 433–34; repr., *Idem, Monetary Change and Economic History in the Medieval Muslim World*, Variorum, 1992. A similar process adopted in the Venetian mint (*zecca*) is described in Frederic C. Lane and Reinhold C. Mueller, *Money and Banking in Medieval and Renaissance Venice*, Baltimore: Johns Hopkins University Press, 1985, pp. 150–51.

39. *Ain-i Akbari*, I, p. 14; tr. p. 19.
40. *Ain-i Akbari*, I, pp. 14–15; tr. pp. 19–20. The preparation of gold beyond 6.5 *ban* was done by a separate method.
41. Sarma, 'Varnamalika System', p. 385.
42. The bullion and coins on sale in the seventeenth century were assayed to the highest degree of only 10 *ban*. *Dastur ul Amal-i Alamgiri*, f. 76b; Bardi Ram Kayath, *Dastur ul Amal-i Mamalik-i Mahrusa-i Hindustan*, OIOC, MS. Or. 1840, f. 136b; *Selected Waqai of the Deccan, (1660–1671 AD)*, Yusuf Hussain (ed.), Hyderabad, 1953, p. 36. In an administrative manual written in 1679 AD, gold of the highest degree of purity (*awwal awwal*) was still reckoned at the standard of 12 *ban* followed by 10 *ban* (*awwal*). Jagat Rai Shujai Kayath Saksena, *Farhang-i Kardani*, MS Aligarh, Maulana Azad Library, Aligarh Muslim University, Abdus Salam, Farsiya 315/85, f. 15a.
43. Tavernier, *Travels in India*, I, pp. 28–29; *Factory Records, Dacca*, 1, f. 80b; *Factory Records, Surat*, 5, ff. 170a–b.
44. *Hidayat ul Qawaid*, f. 37b; *Kaghzat-i Mutafarriqa*, ff. 57a–58b; *Document Forms*, ff. 233b–234a; tr. p. 66; *Mirat-i Ahmadi, Supplement*, p. 183.
45. *Kaghzat-I Mutafarriqa*, ff. 58a–b; *Hidayat ul Qawaid*, f. 38b. Raw silver (*kham nuqra*) and certain types of coins were first thrashed and pulverised in a separate department (*sudakhana*).
46. *Ain-i Akbari*, I, pp., 17–19.
47. *Hidayat ul Qawaid*, ff. 37b, 38a–38b.
48. *Ain-i Akbari*, I, pp. 13–14; tr. pp. 18–19.
49. 'Experts of this art now tell wonderful stories (about the purity of gold) and attribute it to fairy tales (*afsana*) and alchemy (*kimiya*)'. Ibid.
50. Tavernier, *Travels in India*, I, pp. 8–9.
51. *Ain-i Akbari*, I, pp. 16–23.
52. For the use of *pachak-i dashti* as fuel to refine gold see *Risala dar bab kushtan Jami-i Rasayan*, Bodl. MS. Fraser 205, f. 36a. In the *Ain* and *Persian-Urdu Vocabulary* (op. cit., f. 25) the term *upla* is also mentioned.
53. In the margin of an MS consulted by Blochmann (*Ain-i Akbari*, I, p. 18, f. n. 7) *ruhama*, *dama* and *dhunkni* appear as names of the instrument made of animal skin and used by ironsmiths and goldsmiths to blow fire.

54. On account of the labour and skills involved in forging between 170 and 200 plates for each gold consignment of 100 *tolas*, the plate makers received the highest wages – 43 per cent of the total – paid in the department of gold minting.
55. The fraudulence apparently constituted substitution of deficient gold, already evaluated for its mint price, during the process of refining.
56. *Ain-i Akbari*, I, p. 17; tr. p. 22. In one of the earliest descriptions of the method of refining gold in the mints of the Central Islamic Lands, the gold plates were heated with salt, sulphur and brick powder in the furnace. One of the plates was tried on the touchstone and the whole mixture could be reheated. Abu Muhammad al Hasan b. Ahmad al Hamadani, *Kitab al Jauharatain al Atiqatain al Maitam min al Sarfa wal Baida* (Southern Arabia, first half of the 10th Century); extracts translated in Christopher Toll, 'Minting Techniques according to Arabic literary sources', *Orientalia Suecana*, xix–xx (1970–71), p. 130. According to the *Rutbat al Hakim* (Arab Spain, c. 1000 AD), gold leaves were heated with alum, salt and finely powdered baked clay. See E. J. Holmyard, 'Maslama al Majriti and the Rutbat al Hakim', *Isis*, VI, iii (1924), pp. 304–5. In yet another description (in Ibn Bara's *Kashf al-asrar*) where the method was called *taliq*, the gold ore was melted, cast into ingots and made into thin plates. The plates were heated with the above mixture, called *turab at taliq*, in cups sealed with clay in a furnace for one night. Ehrenkreutz, 'Extracts from the Technical Manual', p. 428; Toll, 'Minting Techniques', p. 130. In Abul Hasan Ali b Yusuf al Hakim's, *Ad-Daula al mustabika fi dawabit Dar as Sikka* (Morocco, c. 1336–72 AD) the method is still the same, though it was called *tashhir*. Abul Hasan also mentions the physical examination of the refined gold plates before they were tested on the touchstone. Toll, 'Minting Techniques', pp. 135–36.
57. *Ain-i Akbari*, I, pp. 19–20, 31–32; *Home Miscellaneous*, H/357, f. 430 ('Nearahs or persons who wash and sift the Dross'). Also see, Zafar ur Rahman Dehlawi, *Farhang-i Istilahat-i Peshawaran: A Glossary of Technical Terms Used in Indian Arts and Crafts* (8 Vols., Delhi, 1939–44), IV, s.v. *niyariya*.
58. *Punhar* was prepared by heating lead and a measured quantity of the ashes of the *babul*-wood (six fingers for every *man* of lead) with charcoal. After the lead melted, the coals were removed and the furnace was covered with bricks till the ashes soaked up the lead.
59. The acid (*tezab*) was made from potassium hydrochloride (Hindi *sajji*; Pers. *ashkhar*) and saltpetre. Cf. Iqbal Ghani Khan, 'Metallurgy in Medieval India – 16th to 18th Centuries', *The Technology in Ancient and Medieval India*, Aniruddha Roy and S.K. Bagchi (eds.), Delhi: Sudeep Prakashan, 1986, p. 83.
60. *Ain-i Akbari*, I, pp. 18–19.

61. Ibid., p. 19. The author of *Tadhkirat ul Muluk* (f. 37a) mentions that the silver reached perfect fineness (*kamil aiyar*) when *shakhchas* appeared in the shape of bubbles (*ba shakl-i hubab*) on the surface of the bar. He further adds that pure silver was therefore called *shakhdar*. In an eighteenth century Persian dictionary, *shakhdar* is described as the purest form of silver called *chandi* in Hindi. Anand Ram 'Mukhlis', *Mirat-i Istilah*, OIOC, MS Or. 1813, f. 173a.
62. In the eighteenth and nineteenth centuries, the technique used to refine silver was pretty much the same. See, Om Prakash, 'Foreign Merchants', p. 176; John Malcolm, *A Memoir of Central India*. 2 Vols. London: Kingsburry, Parbury and Allen, 1824, II, p. 82 fn. Cupellation was the standard method used in the middle ages to refine silver. See R.J. Forbes, 'Metallurgy' in *A History of Technology*. Vol. II, *The Mediterranean Civilization and the Middle Ages c. 700 BC to c. 1500 AD*, Charles Singer, E.J. Holmyard, A.R. Hall and Trevor I. Williams (eds.), Oxford: Clarendon Press, 1956, p. 42; Holmyard, 'Rutbat al Hakim', p. 304; Toll, 'Minting Technique', pp. 133–36.
63. *Home Miscellaneous*, H/357, ff. 429–430. Traces of lead however remained in the refined silver as can be seen from the figures in the final column of Table 5.
64. *Ain-i Akbari*, I, p. 22.
65. In the late seventeenth century, the loss of gold in refining the Japanese coin (*koban*) of just about the same fineness as the gold of 8.25 *ban* was 14.28 per cent, whereas the quantity recovered was 2.64 per cent. *Beschryvinge*, I, ii, p. 92. Although the figures calculated from the Dutch account of the Rajmahal mint are in close proximity with those of the late sixteenth century, the quantity of gold lost and recovered would vary with the fineness of the metal and the technique employed for refining.
66. *Ain-i Akbari*, I, pp. 31–32.
67. Ibid., pp. 22–23; *Mirat-i Ahmadi*, I, p. 340.
68. *Ain-i Akbari*, I, p. 19; *Document Forms*, f. 223b; *Hidayatul Qawaid*, f. 38b. The record of Calcutta mint calls the *chashnigir* ('Chockneyker') a 'native assayer'. *Home Miscellaneous*, H/357, ff. 429–30.
69. Malcolm, *Memoir*, II, p. 82 fn. See *Hidayat ul Qawaid*, f. 39a for the instruction regarding the right alloy composition in order to avoid the damage done to the edges of the coins while striking.
70. Abul Fazl does not directly mention the addition of alloy in the refined gold and silver. In the *Ain*, the weight of the total number of rupees struck (1006) was exactly the same as the net weight of the ready-to-be-minted silver (964 *tolas*). In the seventeenth century, the general impression was that the rupees were 'made of perfect silver without an alloy'. Edward Terry, *Early Travels in India, 1583–1619*, William Foster (ed.), London: Oxford University Press, 1921, p. 302. *Ain-i Akbari*, I, p. 19.

71. The analysis was done by M.R. Cowell et al., Department of Coins and Medals, British Museum..
72. Haider, 'Monetary Integration of India', pp. 137–40.
73. *Ain-i Akbari*, I, pp. 16–17.
74. Ibid., *Ain-i Akbari*, I, pp. 18–19, 32–33; *Dastur ul Amal-i Alamgiri*, ff. 54a, 55b, 56b, 57a; *Hidayat ul Qawaid*, f. 39b. A reduction in the loss of weight would suggest some improvement in the technique of either refining or casting employed in seventeenth century Mughal mints. In the Safavid mint of Isfahan, the loss in casting refined silver was from 0.66 to 1 per cent. *Tadhkirat ul Muluk*, f. 37b. In the medieval Venetian mints the figure was 0.4 per cent. Lane and Mueller, op. cit., p. 215.
75. Memoir, II, p. 82.
76. *Ain-i Akbari*, I, pp. 17–18; *Kaghzat-i Mutafarriqa*, ff. 57b–58b; *Hidayat ul Qawaid*, f. 38a; *Beschryvinge*, III, ii, p. 109 (Appendix VIIIc).
77. *Kaghzat-i Mutafarriqa*, ff. 58a–b. In the Calcutta mint, the *zarrab* was aided by seven persons in his work of cutting blanks, weighing, shaping and polishing. *Home Miscellaneous*, H/357, ff. 467–69.
78. *Ain-i Akbari*, I, p. 17. The edited text has *sandan* (anvil) but *mizan*, found in three manuscripts, suits the context better. The process used in the Isfahan mint for cutting blanks was called *qattai* (lit. cutting). *Tadhkirat ul Muluk*, f. 36a.
79. *Memoir*, II, pp. 82–83. The blanks were boiled ('pickled') in a mixture of tamarind and salt (*namak wa imli*) and cleaned and polished before being sent for stamping. *Hidayat ul Qawaid*, ff. 37a, 39b; Malcolm, *Memoir*, II, p. 83 fn.
80. *Ain-i Akbari*, I, pp. 17, 32–33.
81. *Hidayat ul Qawaid*, f. 38a. The MSS have *farsh ra bar musala dashta* and (the variant) *farsh ra musala wa dasta*. Could the expression be *farsh ra dar musala wa dasta dashta* ('the blank was placed between the two dies')?
82. *Ain-i Akbari*, I, pp. 17–18; *John Marshall in India*, p. 117.
83. The configuration was not always binary and principal legends could appear on both sides of the coin (for instance, *kalima* and caliph on one side and Jalaluddin Muhammad Akbar Badshah Ghazi on the other) to render the nomenclature subjective.
84. *Ain-i Akbari*, I, pp. 17–18, 23, 47–48. In the Mughal Empire, the post of the engraver was usually reserved for those well versed in reading and writing Arabic and Persian. In a Dutch record of minting operation in Gujarat in the seventeenth century, the engraver is described as '*moleh of priester, die de letters op de stampel snijdt*' ('the *mulla* or priest, who carves letters on the die'). *Beschryvinge*, II, iii, p. 108.
85. Helen W. Brown, 'The Medieval Mint of Cairo: Some Aspects of Mint Organisation and Administration', *Later Medieval Mints: Organisation, Administration and Techniques. The Eighth Oxford Symposium on Coinage and*

Monetary History, N.J. Mayhew and Peter Spufford (eds.), Oxford: Oxford University Press, 1988, pp. 32–33. For the centralised distribution and withdrawal of dies in the Ottoman mints see *Al-Damurdashi's Chronicle of Egypt: 1688–1755. Al-Durra al-Musana fi Akhbar al-Kinana*, tr. and annotated, Daniel Crecelius and Abd al Wahhab Bakr, Leiden, 1991, p. 40.

86. Tavernier, *Travels*, I, p. 25. Although there was a law against counterfeiting based on the provisions of the *sharia* (*Mirat-i Ahmadi*, I, p. 280), its implementation seems to have been left to the discretion of local officials. See *Selected Waqai of the Deccan*, p. 107; Fryer John, *A New Account of East India and Persia being Nine Years' Travels 1672–81*, 3 Vols., W. Crooke (ed.), London: Hakluyt Society, 1909–15, I, p. 244.

87. A clause in the treaty dictated by the Mughal Emperor, Shahjahan, to the ruler of Golconda stipulated that Mughal silver and gold coins should be struck by with the dies sent from the imperial court (*dargah-i alam panah*). Abdul Hamid Lahori, *Badshahnama*, 2 Vols., 3 pts., Abdur Rahim and Kabiruddin Ahmad (eds.), (Bib. Indica, 1867–68), I, ii, p. 178.

88. *Ain-i Akbari*, I, p. 18; *Kaghzat-i Mutafarriqa*, ff. 57b, 58a–b; *Hidayat ul Qawaid*, ff. 38a, 39b; *Beschryvinge*, II, iii, pp. 108–09; *John Marshall in India*, p. 117. Malcolm's description is once again detailed and instructive. The blanks were placed between the two steel dies, the pile being firmly fixed in an anvil and the upper die, coming over it in the form of a 'large heavy punch', was struck with a heavy hammer. One blow sufficed for the inscriptions to be stamped. The two men were relieved every two hours. *Memoir*, II, p. 82. In a late eighteenth century painting at the Ashmolean Museum, Oxford, which shows coining at the Calcutta mint, the hammerer levels himself to a reasonable height of the punch by standing in a hole. The painting is reproduced in Dennis R. Cooper, *The Art and Craft of Coinmaking: A History of Minting Technology*, London: Spink & Son, 1988, p. 26. A similar scene from the mint is depicted in another (British Museum) Mughal painting reproduced in *Money: A History*, Jonathan Williams (ed.), London: British Museum Press, 1997, p. 121 (Illustration 3).

89. *Hidayat ul Qawaid*, f. 38b (*marammat-i ghan wa musala*). Sometimes, the supplier was charged for the repair of the hammer (*'verstellen der hamers van den seraab ofte muntslager'*). *Beschryvinge*, III, ii, p. 109 (Appendix VIIIc). The Calcutta mint employed blacksmiths (*lohars*) 'for making and repairing the mint utensils'. *Home Miscellaneous*, H/357, ff. 429–430.

90. *Ain-i Akbari*, I, p. 18; *Beschryvinge*, II, iii, pp. 108–09; *Kaghzat-i Mutafarriqa*, ff, 57b–58b.

91. A mechanical process of production was introduced in the Ottoman Empire towards the end of the seventeenth century as a result of which the daily production of copper coins in the Istanbul mint rose to very high proportions. Halil Sahillioglu, 'Introduction of Machinery in the Ottoman Mint', *Transfer of Modern Science and Technology to the Muslim World*, Ekmeleddin Ihsanoglu (ed.), Istanbul: Research Centre for Islamic

History, Art and Culture, 1992, pp. 262–66. In contrast, European mints were set for mechanical production at the turn of the century. The Florentine mint relied solely on hand technology till 1576, when an extraordinary influx of American silver necessitated the use of hydraulic power (water mill) in the German fashion. The old mints, however, continued to strike gold coins for quite some time. The introduction of new technology increased the output by 73 per cent. Carlo M. Cipolla, *Money in Sixteenth Century Florence*, Berkeley: University of California Press, 1989, pp. 90–92.

14 Massive Forge-Welded Iron Cannons of India

R. Balasubramaniam

The high status of iron and steel technology in ancient and medieval India is amply reflected in the manufacture and use of numerous large iron objects. One of the significant large iron objects manufactured in medieval India was forge-welded cannon (Schwarz, 1901; Neogi, 1914). The massive forge-welded iron cannons located at different parts of India are living testimonies to the mastery of the Indian blacksmiths in the technology of extraction of iron and its forging into large shapes. In spite of their historical and technical importance, the forge-welded iron cannons of the Indian subcontinent have not been systematically and thoroughly studied. While the history of these cannons is practically unknown, more conspicuous by absence are technical literatures pertaining to their design, construction and use. It is very important to address these cannons because these are shining examples of the medieval Indian blacksmith's skill in the design, engineering and construction of large forge-welded iron objects. When compared with forge-welded cannons available in the rest of the world, the Indian specimens possess certain unique design features. The relatively large number of existing Indian forge-welded iron cannons is also a point to be noted. It must be noted that, in this article, we specifically refer to cannons used in land warfare. It is known that forge-welded iron cannons, known as 'port-pieces' were deployed in European ships of the 16th century (Smith, 1993) and it is probable such cannons were also employed in Indian ships (Khan, 2004a). It must be emphasised that when we are referring to the term 'medieval', it is really the period from about AD 1500 to AD 1800. Therefore, this period can also be called the 'pre-modern' period.

Europeans pioneered the manufacture of cast iron cannons and most of the cast iron cannons that are located in Indian forts can be traced to European influence. This paper will not address cannons manufactured by the cast iron technology. Interestingly, the Chinese appear to have mastered casting of monumental iron objects quite early in history (Wagner, 2000), at least since the eighth century, as evidenced, for example, by the recently discovered magnificent 'iron oxen' and 'iron men' of Pujin. The massive Pujin castings were cast in AD 724 and served as anchor-weights for the Pujin Bridge, which crossed the Yellow River outside the ancient city of Puzhou in modern Yongji County, Shanxi (Paludan, 1994).

There are scattered reports on the many large cannons present in different parts of India (Schwarz, 1901; Neogi, 1914; Roessler, 1997; Kannan and Balasubramaniam, 2000; Khan, 2001; Chatterji, 2001; Kedareswari, 2005). The author has undertaken detailed studies of some of the forge-welded iron cannons (Balasubramaniam, 2005a, 2005b; Balasubramaniam et al., 2004; 2005a, 2005b, 2005c, 2005d, 2005e; Neff and Balasubramaniam, 2005). In addition, a unique type of cannon composed of forged iron interior and cast bronze exterior has been studied (Balasubramaniam, 2005c) and its similarities with massive bronze cannons (Balasubramaniam, 2005d) noted. A brief description of these cannons will be provided. Other prominent massive forge-welded cannons located in the Indian subcontinent will be presented. The wrought iron cannons found in different parts of India were basically manufactured by hooping iron rings over longitudinally placed iron staves. The medieval blacksmiths continued the rich tradition of manufacturing iron objects by the forge welding technique, which was the method used to fabricate small and large iron objects like the Delhi (Balasubramaniam, 2002a) and Dhar (Balasubramaniam, 2002b) iron pillars.

Panchanan Neogi (1914), in his landmark book *Iron in Ancient India*, provided the first catalogue of some of the wonderful forge-welded iron cannons of India. He followed in the footsteps of P.C. Ray, who collected a large amount of medical and medico-chemical Sanskrit literature bearing on the knowledge of ancient Indians on metals and their compounds in his well-known *History of Hindu Chemistry* (Ray, 1909). Considering the time period in which the book was published, it must be appreciated that Neogi was the first person to recognise the importance and significance of these wonderful pieces of Indian engineering skill and

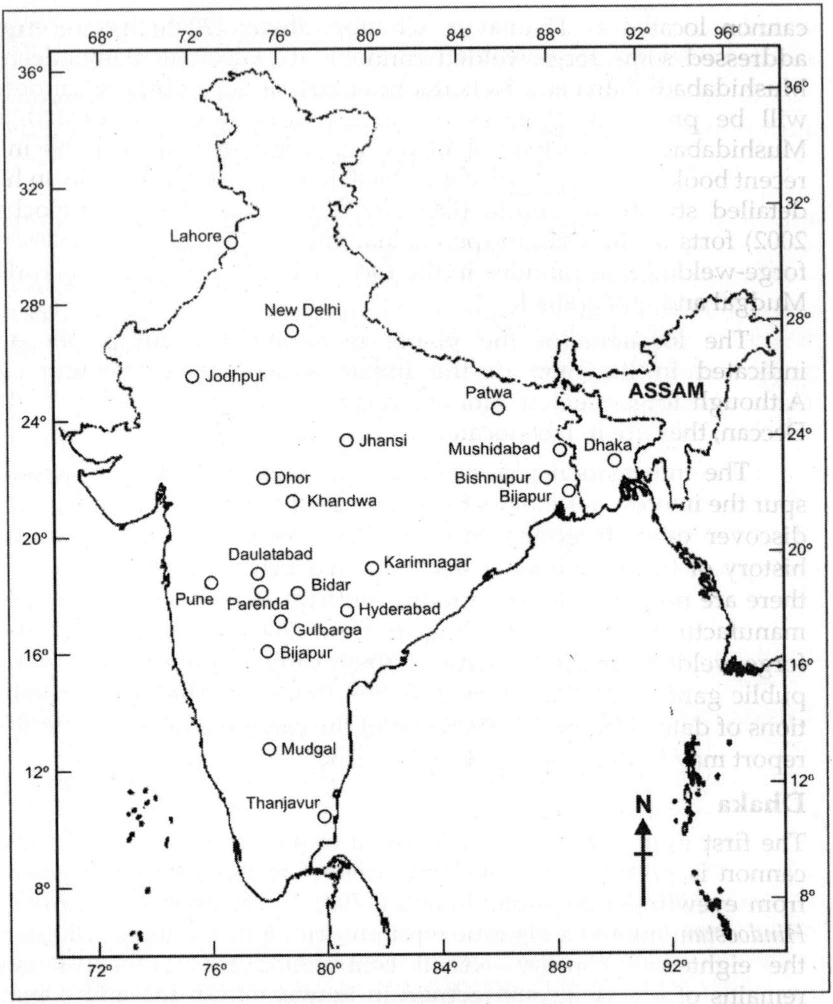

Figure 1: Map of the Indian subcontinent showing the location of places referred in the present paper.

systematically catalogue them. His aim was to highlight the skill of the Indian blacksmiths and provide evidences for their skill in fabricating large iron objects. Neogi described some notable forge-welded cannons in Dhaka, Mushidabad, Bishnupur and Bijapur. The author has not studied most of these cannons, yet. The author has personally studied the forge-welded iron cannons at Thanjavur, Bishnupur, Hyderabad (Golconda Fort and Gun Foundry), and Jhansi (fort). Roessler (1997) has also studied the cannon located at Thanjavur. Chattopadhyay (2005) has recently addressed some forge-welded cannons at Dhaka (in Bangladesh), Mushidabad, Patna and Kolkata. Brief descriptions of these cannons will be provided. Pictures of forge-welded cannons located at Mushidabad, Bidar, Hyderabad, Jodhpur and Pune are available in a recent book on vintage guns of India (Chatterji, 2001). Deloche, in his detailed studies of Hindu (Deloche, 2001) and Muslim (Deloche, 2002) forts of the Deccan (peninsular India) has identified massive forge-welded iron cannons in the forts of Bidar, Gulbarga, Parenda, Mudgal and Hyderabad.

The locations of the places mentioned in this paper are indicated in the map of the Indian subcontinent in Figure 1. Although forge-welded cannons appear to be concentrated in the Deccan, they are in fact, located all over the country.

The information presented in this paper, it is hoped, should spur the interest of scholars to take up study of these cannons and to discover other forgotten forge-welded cannons. Research on the history of these cannons is also required because, in several cases, there are no available inscriptions on them to indicate their precise manufacturing date. According to Khan (2004a), the earliest dated forge-welded cannons are two wrought-iron cannons lying in the public gardens at Khandwa (Madhya Pradesh). They bear inscriptions of dates 1585 and 1589. Some of the cannons catalogued in this report may be dated earlier to this period.

Dhaka

The first example to be quoted is interesting because this massive cannon is presumably lost. Some details of this cannon is known from eyewitness accounts. Renell (1793), in his *Memoir of a Map of Hindoostan*, quoted a gigantic wrought-iron gun located at Dhaka in the eighteenth century. Renell commented that "there are the remains of a very strong fortress in it; and within these few years there was near it a cannon of extraordinary weight and dimensions: but it has since fallen into the river, together with the bank on which it rested". Luckily, Renell took measurements of the gun very carefully. The length of the cannon was 22′ 10.5″, diameter at the breech 3′ 3″, diameter at 4 feet from the muzzle 2′ 10″, diameter of the muzzle 2′ 2″, and diameter of the bore 1′ 3″. Renell further quoted that this massive cannon "was made of hammered iron; it being an immense tube formed of 14 bars, with rings of 2 or 3 inches wide driven over them, and hammered down into a smooth surface, so that its appearance was equal to that of the best executed pieced of brass ordnance, although its proportions were faulty. The gun contained 2,34,413 cubic inches of wrought iron and consequently weighed 64,814 pounds avoirdupois (about 30 tons), or about the weight of eleven 32-pounders. Weight of an iron shot from the gun, 465 pounds".

Rennel has noted that this cannon has sunk in the Ganga. The location of this cannon is presently not known. It is important to trace this cannon, as this is a fairly massive cannon. If proper investigations are conducted, it is quite possible to retrieve this cannon. Quoting from local sources, Rennel stated that the name of the cannon was *Kaley Khan*, and that a cannon bearing the same name was mentioned by the Venetian traveller Manucci (1906) in the list of big cannons possessed by the Mughals in the seventeenth century.

Figure 2a

Figure 2b

Figure 2: Large forge-welded cannon in Dacca: (a) located at Osman Udhan (Garden) (Photograph courtesy: Pranab Chattopadhyay), and (b) old photograph showing its original location near the river.

Neogi (1914) also mentioned, that "a smaller iron gun still exists at Dacca which was constructed at the same time as the bigger one. Mr H.E. Stapleton took its measurements, the gun being 11 feet long and the diameter of its mouth 1 foot 7½ inches".

Chattopadhyay (2005) has identified a large forge welded cannon in Dacca, which is standing in Osman *Udhan* (Garden). A picture of this cannon is provided in Figure 2a. An old photograph of the same is provided in Figure 2b. The history of this cannon is not clear. The name of this cannon has been mentioned *Kalu Jham Jham* and also as *Bibi Mariyam*. Its total length is 538 cm. Four rings were originally attached to the clamps.

Mushidabad

A large cannon at Mushidabad (Figure 3a) was first described by Neogi (1914) as follows: "The great iron gun at Mushidabad, called *Jahankosha* (Conqueror of the World), 17 feet 6 inches in length and with a circumference of 5 feet 3 inches, is of great historical importance owing to the existence of nine inscribed brass plates on it. From the inscription it is ascertained that this gun was constructed in 1047 Hijira, Jamadinssani (October, 1637), during the reign of Emperor Shajahan (the then Subedar of Bengal being Islam Khan) at Jahangirnagar (Dacca) by chief mechanic Janardan under the supervision of Daroga Sher Mahammad and Inspector Harabullava Das." Neogi noted, generally, that the cannons "were all manufactured by

Figure 3a

Hindu mechanics and metallurgists under the orders of the Mohammedan rulers. It appears that the Hindus were skilled workers in iron in all ages and their remarkable skill was utilised by the Mohammedan rulers in the manufacture of wrought-iron guns which must be really regarded as specimens of Hindu art". This cannon is now protected by the Archaeological Survey of India (ASI) and placed on a platform in the city (Figure 3b). The excellent forge-welding of this cannon has been noted by Chattopadhyay (2005). The total length is 530 cm and its bore diameter is 15 cm. The estimated weight of the cannon is about 8 tons.

Figure 3b

Figure 3: (a) Iron cannon at Mushidabad as first reported by Neogi (1914). The name of this cannon is *Jahan Kosha.* (b) Current location of this cannon (Chatterji, 2001).

Another massive forge-welded cannon at Mushidabad is the *Bachhawali* cannon (Figure 4a). This is currently located at the garden of the Hazarduari Palace museum. Its construction is most unique in that the powder chamber is of smaller diameter than the shot chamber, very much reminiscent of the massive forge-welded cannons of Europe, like *Dulle Griet* at Ghent and *Mons Meg* at Edinburgh (Ritter, 1938; Smith and Brown, 1989). The total length of the *Bachhawali* cannon is 461 cm, with the length of the powder chamber being 110 cm. The external diameters of the stone chamber and powder chamber are 68 and 42 cms, respectively. The bore diameter is 46 cm. The stone chamber displays ten raised rings, with the first one from the muzzle exhibiting a unique design. Holding rings are provided at the rear end of the *Bachhawali* as well as at the location where the smaller diameter chamber joins the larger diameter chamber (Figure 4b). Notice that at this location, rings are provided both on the smaller as well as larger diameter chambers. These rings are 27 cm in diameter and 2.4 cm in thickness. These rings may be specifically for handling the cannon and not necessarily for screwing-unscrewing the two different diameter sections of the cannon. The earlier misconception that such a mechanism may be operative in the European examples has been conclusively proven to be false based on X-ray radiography studies of *Mons Meg* (Smith and Brown, 1989).

Figure 4a

Figure 4b

Figure 4: (a) The *Bachhawali* cannon of Mushidabad (Chatterji, 2001), and (b) the location where the small diameter chamber joins the larger diameter chamber

Bijapur

The first cannon at Bijapur described by Neogi (1914) was the *Landa Kesab*. He writes, "on the south wall of the city of Bijapur there is still to seen the celebrated *Landa Kesab* gun (Figure 5). It is also made of circular iron rings shrunk on longitudinal iron bars and then welded together. Aurangzeb is credited to have removed the original points on which it turned in order to make it practically useless. It is 21 feet 7 inches long, diameter at the breech is 4 feet 4 inches, caliber 1 foot 7 inches, length of bore 18 feet 7 inches and the estimated weight 47 tons". This cannon has also been described by Cousens (1916). Cousens also described a smaller cannon beside *Landa Kesab* cannon and this can also be seen in Figure 4 and assumed it to be a mortar. Deloche (2005a) has dated the *Landa Kesab* to 1646.

Figure 5: The *Landa Kesab* gun at Bijapur as reported by Neogi (1914). This photograph was reproduced from *The Empress,* May issue, 1913, No. 2, p. 12.

The second cannon described by Neogi was the cannon on top of the Haider Burj (Figure 6). He comments, "the long gun, named the *farflier,* which is said to have been raised on the huge tower known as the Haider Burj at Bijapur by means of an inclined plane over a mile long, should be of interest in this connection. This gun, too, was constructed of iron bars of square section and laid longitudinally along the bore, over which rings were slipped, one at a time, while the last was re-heat. As they cooled they shrank and bound the iron bars together. The gun near the breech was strengthened by the addition of a second layer. It is 30 feet 8 inches long and has a bore 1 foot in diameter". This name of this cannon is *Lambacharri*. Deloche (2005a) provides a date of 1583 for this cannon. He has also mentioned about another wrought iron cannon on the same Haidar Burj of length 604 cm and bore diameter of 20.3 cm.

The present condition of these cannons deserves attention.

Figure 6: The *Lambacharri* (*Farflier*) cannon at Bijapur reproduced from Cousens (1916).

Bidar

There are several wonderful massive bronze and iron cannons located in the Bidar fort. Some of them have been described by Yazdani (1947). Interestingly, he does not mention the material of construction of the major cannons that he had located in the fort. These include the *Ali Barid* on Mandu Gate, *Fath Lashkar* on the north west rampart, *Tope-i-Haidari* on Lal Burj, *Ali Barid* on Kala Burj and a long cannon on the western plateau. Based on the photograph provided by him, it is clear that the long cannon is a wrought iron cannon. Deloche (2005a) mentioned that this cannon is 883 cm long, diameter at the muzzle is 53 cm and diameter of the bore is 15.2 cm. The condition of this wonderful cannon and the material of construction of the other cannons have to be investigated. Although Chatterji (2001) has provided pictures of some of these cannons in his book, he does not discuss about their design features and the material of construction. Some of the cannons from the Bidar fort have also been shifted to other places and one notable example is the forge-welded cannon that is presently in Campal, Goa that originated from Bidar. Its length is 376 cm and the diameter of the bore is 41 cm (Chatterji, 2001).

Gulbarga

There are several forge-welded cannons located on the Gulbarga fort rampart. The largest one was commented upon by Neogi (1914) as follows: "The long gun at Gulbarga seems to have been made in a similar way, but it is worthy of notice that there is double row of iron rings, ten on each side."

One of the significance of this cannon is that it still rests on the pivot and this provides valuable information about the actual operation of forge-welded cannons containing trunnions. This has been discussed by Deloche (2005a). The trunnions of the cannons were connected to a forked swivel, constructed usually of wrought iron. This swivel rested on an iron pivot, which was firmly fixed to the ground in an iron socket. The cannon could then be turned around to face different directions by swiveling over the iron pivot. A heavy segmental recoil wall was generally built at a very short distance from the back of the cannon. This is a common feature noticed in Indian forts when cannons were placed on platforms atop bastions. The wall neutralised the recoil of the cannon when it was fired and thereby prevented stressing of the swivel and pivot. Moreover, there must have been equipment in front of the cannon for resting the cannon and also for aiding its movement for positioning purposes. This was usually a wooden arrangement moving on a grooved surface on the platform (Deloche, 2005a). In this way, the cannon's direction of fire and height could be adjusted.

Thanjavur

Based on its weight and size, the forge-welded iron cannon at Thanjavur must be regarded as one of the massive forge-welded iron cannons in the world. The cannon was manufactured in Thanjavur during the regime of Raghunatha Nayak (AD 1600–1645) (Balasubramaniam, 1994). The Thanjavur cannon formed an important component of the fortification scheme and it was located at the eastern entrance of the ancient fortified city. The cannon is referred to as *Rajagopala*, according to local traditions. The complete history of the cannon has been described, elsewhere (Balasubramaniam et al., 2004; 2005a).

The Thanjavur iron cannon rests on three concrete supports. The iron cannon is totally 751.5 cm in length from end to mouth, including the 31.5 cm projection at the end of the barrel. The outside and inner diameters of the gun barrel are 93 and 63 cms, respectively. All the dimensions in this communication, from now onwards, will be in centimetres. Assuming the hollow cylinder of the cannon barrel extends to the complete end of the cylindrical barrel (length of 720 cm), the minimum weight of the cannon estimated from the known thickness of the barrel (i.e., 15 cm) is 20.6 tons. Including the solid end portion, the minimum weight of the cannon adds to the staggering figure of over 22 tons. Further technical details and dimensions are available elsewhere (Balasubramaniam, 2005b). All holding clamp rings, but for two, are currently missing. Only two of the original eight rings still remain. Similar iron rings can be noted on several other large forge-welded cannons. The rear end of the cannon is made of successively smaller diameter rings.

The handling rings seen on several forge-welded iron cannons were used to adjust the direction of cannon fire. Unlike the cannons with trunnions, these types of cannons were usually placed on massive wooden logs. For example, old photographs of the largest bronze cannon at Bijapur, by the name *Malik Maidan*, shows it resting on large wooden logs (Cousens, 1916). This all-bronze cannon was cast by a Turkish engineer (Mohamed bin Hasan Rumi) for a Deccani king in 1548. Incidentally, this is one of the largest cannons in India. Its length is 426 cm, muzzle diameter is 149 cm and the bore diameter is 71.2 cm. Its estimated weight is 55 tons.

Bishnupur

The cannons at Bishnupur were first mentioned by Neogi (1914), who noted that there were several old iron guns at Bishnupur in Bengal and several inscribed Assam guns of the sixteenth and seventeenth centuries. These have been described by Banerji (1909/1911). One massive cannon at Bishnupur has been studied in detail and reported elsewhere (Balasubramaniam et al., 2005b). It is popularly known nowadays as *Dolmodol*, which is a corrupted version of its original name *Dal Mardan* (Crusher of People). The cannon dates to about AD 1650. The cannon was lying in a half-buried condition for long time. In 1918, the governor general of Bengal, Lord Ronaldshay, on a visit to Bishnupur, ordered it to be dug out from the ground and it was then placed on a laterite stone platform. The discoloration of the bottom regions of the cannon is due to the cannon being buried in the ground earlier.

The iron cannon rests on two concrete supports. The total length of the cannon from the front to the extreme rear of the cannon is 3.84 m. The weight of the cannon, estimated based on the dimensions, is 7.9 tons. Other intricate details are provided elsewhere (Balasubramaniam et al., 2005b). There are two protruding solid cylindrical supports (i.e., trunnions) provided halfway along the length of the cannon, which were used for manipulating the cannon's direction and also for aiding its movement and transportation. A total of 189 iron rings have been used in constructing the tapered cylindrical barrel of the iron cannon. The rear of the cannon has been skillfully designed such that the number of rings has been successively reduced in order to finally reach the solid iron cylinder at the extreme end of the cannon. The construction of the rear end and the trunnions indicate the mastery of the blacksmith over his trade.

Jhansi

There are two massive forge-welded cannons located at the Jhansi Fort, which is associated with the heroic deeds of Rani Laxmibai during the First War of Indian Independence (1857). The *Kadak Bijli* is located immediately on the left, when one enters the Jhansi Fort, through its present entrance. The other cannon is located near the Ganesh Temple, slightly in the interior of the fort and this is called *Bhavani Shankar*. Both these cannons are placed on raised concrete supports.

The *Bhavani Shankar* cannon was constructed during the Maratha period in AD 1725. The total length of the cannon from the front to the extreme rear of the cannon is 502 cm. The weight of the cannon, estimated based on the measured dimensions, is 5 tons. The main barrel was manufactured by hooping two layers of iron rings over 15 longitudinally placed iron staves. A minimum number of 125 iron rings were used in the construction of the cannon. More details about this cannon can be found elsewhere (Neff and Balasubramaniam, 2005). The interesting designs of lion and elephant in the front and rear portions of the cannon must be noted.

The *Kadak Bijli* cannon measures 5.5 m in length and weighs about 5 tons. The barrel of the cannon has been constructed by hooping two iron rings over 11 longitudinal iron staves. The iron rings were hooped over such that the second layer of rings closed the gaps between the rings of the first layer. Valuable information about the construction of forge-welded cannons is available in the damaged front and rear portion of the cannon. This cannon has been described in great detail elsewhere (Balasubramaniam, 2005b). Notice that the iron staves progress all the way to the rear of the cannon and that they are thicker in the rear end of the barrel. The excellent engineering of the powder chamber and the design of connection of the iron staves with the rear must be noted.

Hyderabad

The Golconda fort is a majestic monument, which lies on the western outskirts of Hyderabad city in the Indian State of Andhra Pradesh. The construction of the fort in its present form by Mohammed Quli Qutub Shah was started around AD 1525. The Golconda fort is fortified with two lines of defenses. The inner rampart, about 5 km in circumference, protected the palaces and garrison, which were located on the Golconda hill. The outer rampart, surrounding the entire township of Golconda, is about 11 km long, and this was also strongly fortified. At regular intervals, the outer fortified wall provided with *burj*s (bastions). At these bastions, iron and bronze cannons were strategically located. We shall look at three cannons located on these bastions. Two of them are forge-welded iron cannons, while the third a composite (iron+bronze) cannon. After the fort fell into the hands of the Mughals in 1687, they located some of their wonderful cast bronze cannons at strategic locations along the outer fort wall. Therefore, the bronze cannons at Golconda fort are from the time of Mughal occupation, while the iron cannons were present even before the Mughals took over Golconda. The iron cannons are not provided with trunnions, but rather with rings.

A massive forge-welded cannon is located on a bastion called *Bada Burj* (Big Bastion). It was manufactured in AD 1644. The detailed dimensions and design of the cannon have been described elsewhere (Balasubramaniam et al., 2005c). The total length of cannon, as measured from the front end to the back end of the cannon, is 509 cm. The inner and outer diameters of the cannon are 31 and 69 cm, respectively. The conservative weight of the cannon is about 12 tons. The iron cannon was manufactured using three layers of prefabricated iron rings, hooped and forge-welded over 21 longitudinally placed iron staves, which have been flared out in the front end of the cannon. There are two excellent decorative patterns near the front end of the cannon. The iron cannon was provided with four sets of handling clamps in which iron rings were originally located.

The second forge-welded cannon is located near the Fateh Burj bastion. The date of construction of the cannon is not known with certainty. The total length is about 474 cm. The outer diameter of the cannon at the rear end is 86 cm. The inner diameter of the cannon, measured on the front face, is 30 cm. Complete details of the design and dimensions of the cannon are provided elsewhere (Balasubramaniam et al., 2005d). The iron cannon has been provided with ten handling rings, five on each side. The handling rings have been fixed on handling clamps, located at regular intervals along the length of the cannon.

The third cannon is a unique type of cannon, a composite iron-bronze Mughal cannon, named *Azdaha Paikar* located on the Musa Burj. The name of the cannon translates as 'Dragon Body'. Such fancy names were often given to cannons in the Mughal period. Among the several innovations that were implemented in cannon technology in the Indian subcontinent by the Mughals, Marathas, Rajputs and Sikhs, the strengthening of wrought iron cannon barrels by casting them over with bronze (which we may define as a composite cannon) is unique in the annals of cannon technology (Khan, 2004b). European visitors admired these composite cannons, particularly the excellent joint between the iron and bronze sections. One commentator notes that, "nothing but the different color of the two metals indicated the joining" (Thorn, 1962). A detailed inscription on the *Azdaha Paikar* cannon indicates that it was constructed in AD 1647. Moreover, the weight of the gunpowder (11.0 kgs) and shot (33.5 kgs) are precisely known. The complete history, dimensions and engineering design of the cannon have been described elsewhere (Balasubramaniam, 2005c). The front face of the cannon clearly indicates that it was manufactured by casting bronze on an iron assembly consisting of shrunk-fit iron rings over cylindrical iron barrel. The inner iron barrel has been constructed by curling over two iron plates. The bore diameter is 24 cm and the overall diameter of the front face is 70 cm. The length of the cylindrical barrel, excluding the back portion, is 414 cm. The back portion adds another 38 cm, thereby providing the overall length of the cannon as 452 cm. Assuming the shrunk-fit iron assembly to proceed till the end of the barrel, the weight of the iron assembly (iron barrel and shrunk-fit rings) has been estimated to be 3.28 tons. The weight of the remaining portions, assuming it to be composed primarily of copper, is 13.72 tons. Therefore, the overall weight of the cannon is 17.00 tons.

A smaller composite cannon located on the top of Golconda fort indicates that iron staves were cast over with bronze. This cannon was fabricated in 1666 and is called *Qila Kusha* (Conqueror of the Fort). Its length is 325 cm, its muzzle diameter 38 cm and bore diameter 18 cm (Balasubramaniam et al., 2005f). The back portion of the cannon indicates a central iron rod at the extreme end of the cannon and smaller iron rods placed regularly along the circumference of the base plate. These iron rods must be related to the manufacturing methodology of the cannon, with the central iron rod acting as a support for the iron assembly inside.

It is interesting to note that all-bronze cannons, similar in design to the *Azdaha Paikar* cannon noted above, were manufactured during the reign of Aurangzeb (1658–1707). This period can be considered as a time when the cannon technology had achieved sophistication based on the technical nature of the bronze cannons. A good example is the *Fath Raihbar* (Guide to Victory) cannon located on the Petla Burj of the Golconda Fort outer rampart. The total length of the cannon is 486 cm. The diameter of the muzzle is 70 cm, bore 26 cm and end of the barrel 84 cm. The total weight of the cannon is 16.59 tons. The engineering design of the cannon has been explained in detail elsewhere (Balasubramaniam, 2005d). Analysis of composition of the material of construction indicates that it was leaded bronze. The relatively large percentage of lead (9–12 wt. %) facilitated casting. This cannon is a wonderful example of the Mughal bronze cannon manufacturing technology prevalent during the time of Aurangzeb. Some other massive bronze cannons of Aurangzeb which are located at Daulatabad, Kolkata and Lahore have been described by Chatterji (2001).

Other Locations

Chattopadhyay (2005) has described two forge-welded iron cannons of the Assamese king Jayadhvaja Simha (1648–1663), one lying at Patna museum (length 270 cm and bore diameter 9 cm) and the other at the Asiatic Society, Kolkata (length 190 cm and bore diameter 6.5 cm).

There are numerous cannon scattered all over the Deccani forts. Deloche (2001, 2002) mentions that in all the forts (like Golconda, Bijapur, Bidar, Gulbarga, Parenda, etc.) that he studied, he found numerous cannons still present in their original condition in their original location. For example, Deloche (2005b) found four large forge welded iron cannon lying in Mudgal fort. Another massive cannon located at Parenda fort. Its length is 509 cm, diameter of the muzzle 75 cm and diameter of the bore 30 cm (Deloche, 2005a).

In all the Deccani forts, the cannons are located on ramparts that are difficult for people to negotiate and view these wonderful cannons. Moreover, the ramparts are not properly maintained and it is difficult to move around with thick outgrowth all around. Therefore, there is great need to visit these forts first hand and record the technical details of the cannons located on the ramparts and within the forts. Interestingly, Jaikishan (2005) has located and described 19 large forge-welded cannons at three remote forts (Ramagiri, Elgandal and Jagtial) in Karimnagar district of Andhra Pradesh. He also mentioned, in a private communication, that he has personally viewed over 400 forge-welded cannons in the remote forts of the Deccan.

Although a great number of forge-welded cannons can be noticed in Deccani forts, there are such types of cannons in other forts located at different parts of India. For example, a large forge-welded cannon located in Mehrangarh fort in Jodhpur (Chatterji, 2001). An inscription on the barrel indicates its name as *Nusrat Laskar* (Glory of the Army). Its length is 386 cm and its caliber 43 cms. Six holding rings and two trunninos are found on its barrel, the rear portion of which has four raised rings.

There is need to survey all the forts of the Indian subcontinent to identify forge-welded cannons.

In several instances, forge-welded cannons can also be found within modern city and town premises. For example, the Archaeology Department of Andhra Pradesh recovered two forge-welded cannons from a remote fort in the state and it was placed in the premises of the Department at Hyderabad. The total length of the larger cannon is 550 cm with the diameter of the muzzle being 40 cm and that of the bore 25 cm (Balasubramaniam et al., 2005e). The other cannon is a typical example of a howitzer, a cannon that was used for throwing cannon balls up into the air and into the fortification. These cannons were in a state of neglect when the author discovered these wonderful cannon pieces, recently. The Department is restoring these wonderful pieces.

Apart from the Deccani forts, it is also important to identify pre-modern forge-welded cannons present in establishments connected with the military and police. For example, a forge-welded cannon of the 17th century, now located at the District Police Headquarters in Pune (Chatterji, 2001). Its length is 305 cm and bore diameter is 10 cm. Four handling rings are provided in addition to the trunnions. Therefore, careful scrutiny of pre-modern cannons preserved in military and police establishments will reveal further forge-welded cannons.

Epilogue

There is great need for scholars from different disciplines to study these wonderful pieces and to highlight the metallurgical aspects of forge-welded cannon technology practised in India. There is only limited technical literature on how the cannons were fabricated (both iron and bronze cannons) and therefore this exercise would help in understanding the history of science and technology, a very important aspect of Indian society during the medieval times. There are several other unresolved questions related to cannon technology in India that need to be resolved like the kind of cannon balls used, the effectiveness of the cannons in breeching fortifications, the estimate of cannon ball travel and thrust based on available cannon dimensions, cannon ball weight and gun powder weight, the manufacture, storage and use of gunpowder, details of gun carriages, handling of the large cannons, procedure adopted for firing the cannons, details of accessory and special complementary equipments, the organisation of the artillery, the depiction of cannons in Indian miniature paintings from all available sources, the cannon technology as practised by different powers (Mughals, Marathas, Sikhs, Rajputs), the European gunners and foundry personal who were in the service of Indian monarchs, the locations of foundries and workshops where the cannons were fabricated, the influence of European cast iron technology on Indian forge-welded cannon technology, and information about forge-welded cannons that were deliberately destroyed by the Europeans. The last point may seem surprising because the European powers took great efforts to destroy the numerous forge-welded cannons from the Indian forts that they captured. Dupliex mentions that an enormous number of forge-welded cannons were taken from Senji fort to Pondicherry and melted there (Deloche, 2000). Therefore, there is great need to study the history, evolution, epigraphy, social impact of cannon technology apart from detailed technical study of cannon manufacture, material characterisation and both destructive and non-destructive testing (like radiography) to determine the inner construction details, like the one conducted on a large forge-welded iron cannon of Europe (Smith and Brown 1989).

Acknowledgements

The author would like to thank the Archaeological Survey of India for permission to study the forge-welded iron cannons. The author also thanks Dr J. Deloche for several exciting and critical discussions, and for sharing valuable insights regarding cannons in Indian fortifications.

References

Balasubramaniam, K. (1994). *Thanjavur (600 AD–1850 AD)* (in Tamil). Thanjavur: Anjana Publications, 312–37.

Balasubramaniam, R. (2002a). *Delhi Iron Pillar: New Insights*. Indian Institute of Advanced Study, New Delhi: Shimla and Aryan Books International.

Balasubramaniam, R. (2002b). "A New Study of the Dhar Iron Pillar", *Indian Journal of History of Science* 37(2): 115–51.

Balasubramaniam, R. (2005a). "On the First Catalogue of Forge-Welded Iron Cannon by Neogi", *Indian Journal of History of Science* 40(3): 259–68.

Balasubramaniam, R. (2005b). "Kadak Bijli: The Forge-Welded Iron Cannon at Jhansi Fort", *Indian Journal of History of Science* 40(3): 371–88.

Balasubramaniam, R. (2005c). "Azdaha Paikar: The Composite Iron-Bronze Cannon at Musa Burj of Golconda Fort", *Indian Journal of History of Science* 40(3): 389–408.

Balasubramaniam, R. (2005d). "Fath Raihbar: The Massive Bronze Cannon at Petla Burj of Golconda Fort", *Indian Journal of History of Science* 40(3): 409–30.

Balasubramaniam, R., A. Saxena, T.R. Anantharaman, S. Rigeur and P. Dillmann. (2004). "A Marvel of Medieval Indian Metallurgy: Thanjavur's Forge-Welded Iron Cannon", *JOM* 56(1): 17–23.

Balasubramaniam, R., A. Saxena and T.R. Anantharaman. (2005a). "Rajagopala: The Massive Iron Cannon at Thanjavur in Tamil Nadu", *Indian Journal of History of Science* 40(3): 269–94.

Balasubramaniam, R., K. Bhattacharya and A.K. Nigam. (2005b). "Dal Mardan: The Medieval Forge-Welded Iron Cannon at Bishnupur", *Indian Journal of History of Science* 40(3): 295–320.

Balasubramaniam, R., M. Surender and S. Sankaran. (2005c). "The Forge-Welded Iron Cannon at Bada Burj of Golconda Fort Rampart", *Indian Journal of History of Science* 40(3): 321–36.

Balasubramaniam, R., S. Sankaran and M. Surender. (2005d). "The Forge-Welded Iron Cannon near Fateh Burj of Golconda Fort Rampart", *Indian Journal of History of Science* 40(3): 337–48.

Balasubramaniam, R., K. Gayathri Subramanyam and J. Kedareswari. (2005e). "The Forge-Welded Cannons at Gun Foundry in Hyderabad", unpublished research.

Balasubramaniam, R., S. Srinivasan and S. Jaikishan. (2005f). "Qila Kusha: The Composite Iron-Bronze Cannon at Golconda Fort", unpublished research.

Banerji, Rakhaldas. (1909/1911). *Journal of the Asiatic Society of Bengal.*

Chatterji, S.K. (2001). *Vintage Guns of India*. New Delhi: Macmillan, 26–126.

Chattopadhyay, P. (2005). "An Introduction to the Cannons of Eastern India", *Indian Journal of History of Science.*

Cousens, H. (1916). *Bijapur and its Architectural Remains, with an Historical Outline of the Adil Shah Dynasty*. Bombay: Government Central Press, Parts III and IV, 29–33.

Deloche, J. (2000). *Senji (Gingi), ville fortifiée du pays tamoul.* Paris/Pondicherry: Pondichery University.

Deloche, J. (2001). Études sur les fortifications de l'Inde: III. La fortification hindoue dans le Sud de l'Inde (VIc-XVIIIc siècle), *Bulletin de L'École Française D'Extrème-Orient* 88: 77–134.

Deloche, J. (2002). Études sur les fortifications de l'Inde: IV. La fortification musulmane dans le Dekkan méridional (XVc-XVIIIc siècle), *Bulletin de L'École Française D'Extrème-Orient* 89: 39–106.

Deloche, J. (2005a). "Gunpowder Artillery and Military Architecture in South India (15th–18th Century)", *Indian Journal of History of Science.*

Deloche, J. (2005b). *Personal Communication*. Four cannons were located at the Mudgal fort located in Lingasugur Taluk of Raichur district in Karnataka state. The first one was on Fateh Darwazah, the second on Ali Burj; both these were long wrought iron cannons. The last two cannons were found inside the walls, north of Kati Darwazah, "a short one of large diameter and a long one surrounded by strange rings kept in position by hooks".

Jaikishan, S. (2005). "Forge-Welded Cannons in the Forts of Karimnagar District in Andhra Pradesh State", *Indian Journal of History of Science*.

Kannan, R. and R. Balasubramanian. (2000). *Documentation on the Cannons in the Government Museum Chennai (Madras)*. Chennai: Government Museum.

Kedareswari, J. (2005). *Catalogue of Cannons in Andhra Pradesh State Museum*. Hyderabad: Andhra Pradesh Archaeology Department Publications.

Khan, I.A. (2001). "Early Use of Cannon and Musket in India: AD 1442–1526", in J.J.L. Gommans and D.H.A. Kolff (eds.), *Warfare and Weaponry in South Asia 1000–1800*. New Delhi: Oxford University Press, 321–36.

Khan, I.A. (2004a). *Gunpowder and Firearms: Warfare in Medieval India*. New Delhi: Oxford University Press, 17–91.

Khan, I.A. (2004b). "Artillery in Mughal India: 1556–1739", in *Gunpowder and Firearms: Warfare in Medieval India*. New Delhi: Oxford University Press, 91–127.

Manucci, N. (1906). *Storia do Mogor*. 4 Vols., *History of the Moghuls 1653–1708*. Vol. II, (W. Irvine's Translation), London: John Murray, p. 365.

Neff, D. and R. Balasubramaniam. (2005). "Bhavani Shankar: The Forge-Welded Iron Cannon at Jhansi Fort", *Indian Journal of History of Science* 40(3): 349–70.

Neogi, P. (1915). *Iron in Ancient India*. Calcutta: Indian Association for Cultivation of Science, 32–40.

Paludan, A. (1994). "The Tang Dynasty Iron Oxen at Pujin Bridge", *Orientations*, May, 61–8.

Ray, P.C. (1909). *History of Hindu Chemistry*. Volumes I and II, Calcutta: The Bengal Chemical & Pharmaceutical Works Limited.

Rennell, J. (1793). *Memoir of a Map of Hindoostan, or the Mogul Empire: With an Examination of some Positions in the Former System of Indian Geography, and some Illustrations of the present one, and a Complete Index of Names to the Map*. Third Edition, London: M. Brown, 61.

Ritter Karl, (1938). Aufbau und Herstellung der schmiedeeisernen Steinbüchsen des Mittelalters, *Technische Mitteilungen Krupp: Technische Berichte*, Fried. Krupp AG, Essen, 6(5): 113–29.

Roessler, K. (1997). "The Big Cannon Pipe at Tanjavur", *Metal News*, 19(1): 1.

Schwarz, C. Ritter von. (1901). "Uber die Eisen und Stahlindustrie Ostindiens (About the Iron and Steel Industry of East India)", *Stahl und Eisen* 21: 209–11 and 277–83.

Smith, R.D. and R.R. Brown. (1989). *Bombards Mons Meg and Her Sisters*. England: Royal Armouries, The Dorset Press.

Smith, R.D. (1993). "Port Pieces: The Use of Wrought Iron Guns in the Sixteenth Century", *Journal of the Ordnance Society* 5: 1–10.

Thorn, W. (1962). *Memoirs of War in India, 1803–06*, 1818, 117 cited by W. Irvine, *The Army of the Indian Moghuls: Its Organisation and Administration*, originally published from Calcutta 1896, reprint, New Delhi: Eurasia Publishing House, 139.

Wagner, D.B. (2000). "Chinese Monumental Iron Castings", *Journal of East Asian Archaeology* 2(3–4): 200–24.

Yazdani, G. (1947). *Bidar: Its History and Monuments*. Oxford: Oxford University Press, 35–43 and 85.

15 Scientific Attitudes in Medieval North India

Context, Nature and Articulation (AD 1000–1500)

Surinder Singh

This paper seeks to assess the nature of scientific attitudes in medieval north India during the pre-Mughal times. The exercise begins with an analysis of Alberuni's comprehensive observations on the Indian sciences in the early eleventh century. Thereafter, we encounter Amir Khusro's gallant attempt at demonstrating the superiority of Indian learning, including scientific knowledge and cultural forms. The Persian chronicles of the fourteenth century enable us to understand the manner in which the process of state formation generated scientific attitudes among several sections, who were constrained to live with a regime of statistics and records. A perusal of devotional poetry, which was produced by bhakti reformers in vernacular languages, also promises to provide insights into the social progress witnessed by northern India prior to the establishment of the Mughal empire.

I

Alberuni has presented a detailed analysis of the state of sciences in medieval north India and has identified the factors responsible for stagnation in the growth of scientific thinking and difficulties in the learning and dissemination of scientific knowledge.[1] Sanskrit, the language used by the learned class, could not perform its function in an appropriate manner. Owing to a peculiar vocabulary, it had many words to denote a single phenomenon and conversely a single word to denote a variety of phenomena. The negligence of Indian scribes prevented the production of the faithful copies of a text, so that the enormous mental effort of the authors was spoiled. Scientific books were composed in particular forms of poetry, as they were meant to be memorised. The learned classes considered "as canonical only that which is known by heart, but not that which exists in writings". The requirements of poetic metre introduced constrained pharaseology and a certain amount of verbosity.[2] The Indian scientific community, owing to self-conceit and lack of interaction with other countries, refused to acknowledge the existence of scientific knowledge other than their own.[3] They were reluctant to share this limited body of knowledge with Indians of another caste and more so with aliens. Whenever they encountered a foreigner possessing some knowledge superior to their own, they regarded him as a sorcerer or "as water which is so acid that vinegar in comparison is sweet".[4]

Alberuni was not only critical about the preservation and communication of scientific knowledge, he was also critical about the content and quality of Indian sciences as such. In his view, the Indians did not have philosophers of the stamp of Socrates who were both capable and willing to bring sciences to a classical perfection. In most cases, their scientific propositions were in a state of utter confusion, devoid of any logical order and mixed up with the silly notions of the crowd and religious dogmas which popular belief did not dare to question. He regarded the literature on mathematics and astronomy as a mixture of pearl shells and sour dates, or pearls and dung, or costly crystals and common pebbles. The Indian scientists did not see any distinction between the two, as they could not conform themselves to the methods of a strictly scientific deduction.[5]

Alberuni was convinced about the stranglehold of theology on the Indian sciences. The Hindu scriptures, including the Puranas, were replete with assertions regarding the shape of the earth, which stood in direct opposition to scientific truths as known to Indian astronomers. A majority of Indians found it essential to consult these books while performing a variety of religious rites and, in the process, remained entangled in astronomical calculations and astrological predictions which could not stand the test of scientific scrutiny. In consequence of this trend, the common masses held their astronomers in high regard. The astronomers, with a view to repay the people for this public adulation, accepted the popular notions as scientific truths and conformed themselves to ordinary standards of scientific knowledge and churned out such spiritual stuff as required by them. With the passage of time, two theories – the vulgar and the scientific – were intermingled. The scientific principles of astronomy were pushed into a state of confusion. In particular, this applied to a majority of astronomers who assimilated the received knowledge and did not subject it to independent scientific research. They also followed the theologians in everything that did not encroach on their sciences. For example, they had adopted the theory of Mount Meru being under the north pole and that of the island of Vadavamukha lying under the south pole.[6]

Moving away from these general observations on the state of Indian sciences, we would narrow down to Alberuni's comments on the understanding of Indian astronomers on the solar and lunar eclipses. Alberuni observes that the Indian astronomers were perfectly familiar with the causes of the solar and lunar eclipse. In his Samhita, Varahamihira has affirmed that the former occurred when the moon covered a part of the sun and the latter occurred when the earth covered a part of the moon, during the course of their movements in the respective orbits. The solar eclipse did not revolve from the east, nor the lunar eclipse from the west. When the sun was approached by moon from the west, the moon covered the sun like a portion of a cloud. The magnitude of this covering differed in different regions. Since that which covered the moon was large, its light waned when half of it was eclipsed. Since that which covered the sun was not large, its rays continued to be powerful in spite of the eclipse. Having underlined the cause of the two eclipses, Varahamihira complained that some scholars (Brahmins or orthodox theologians) and the common people believed that the eclipse was caused by the Head, who was a demon (Rahu) with a body like the sun and moon, but which was dark and black. Some people believed that his body resembled a serpent, while others held that it had no body except a black colour.[7] Whatever be his form, he intruded into the heaven and, appearing among the angels, managed to have a share of the nectar (*amrita*) being distributed by Vishnu. When Vishnu realised his mistake, he delivered a fatal blow on the Head, who did not die as he had consumed the nectar. He was accepted as an inhabitant of heaven, but could appear only on the occasion of the eclipse.

Alberuni has expressed his surprise at Varahamihira, who offered a long argument demolishing the popular notion regarding Head as the cause of the eclipse, but ended up accepting the existence of Head as a celestial body and thus compromised with the traditionalists. Explaining his stance, Varahamihira stated that the Head humiliated itself after being cut off and received from Brahma a portion of the offering which Brahmins give to fire during the eclipse. Therefore, the Head was found near the site of the eclipse, searching his portion of the offering, but he had nothing to do with the cause of the eclipse. In his attempt to understand this contradiction, Alberuni has praised Varahamihira as a man whose knowledge regarding the shape of earth was accurate, whose position stood firmly on the foundation of truth and who also spoke the truth. However, he seemed sometimes to side with the Brahmins, to whom he belonged, and from whom he could not separate himself.[8]

Alberuni has recognised Brahmagupta as the most distinguished of Indian astronomers, but has exposed his unscientific stance regarding the cause of the eclipse. Brahmagupta did not agree with those scientists – Varahamihira, Srishena, Aryabhatta and Vishnuchandra – who held that an eclipse was caused by the shadows of the earth and moon. Instead his belief that the eclipse was caused by the Head was based on three arguments: (i) the generality of inhabitants of the world nurtured the idea, (ii) the Vedas, Manu Smriti and Garga Samhita upheld Head as the cause of the eclipse, and (iii) if the Head did not cause the eclipse, the rituals practised by the Barahmins on the occasion of an eclipse would be rendered illusory and unrewarded by heavenly bliss.[9] While criticising Brahmagupta, Alberuni attributes the basic mistake to Puranic literature which placed moon above the sun, so that any celestial body which was higher could not cover that which was lower in the sight of those who stood lower than both. Therefore, they required a being which devoured the sun and moon, as the fish devoured the bait, causing them to appear in shapes in which their eclipsed parts appeared in reality. Assuming an air of superiority, Alberuni declared that in every nation there were ignorant people and leaders who were even more ignorant than the people.[10]

Alberuni found that Brahmagupta practised science and non-science at the same time. On the one hand, the distinguished Indian astronomer followed the Vedas, Puranas, Manu Smriti and Garga samhita regarding Head as the cause of the eclipse. On the other hand, he sought to calculate the diameter of the moon to explain the solar eclipse and the diameter of the shadow of earth to explain the solar eclipse. Secondly, if Brahmins performed some religious rites on the occasion of an eclipse, the eclipse merely denoted the date of these rituals and not their cause. Thirdly, Brahmagupta could not claim to base his stance on the opinion of the generality of the inhabitable world, as it could not be determined either by exact research or by means of historical traditions. If he had Hindus in mind, the number of people who differed from them, both in religion and law, was larger than those who agreed with them.

Among the Hindus, the educated were outnumbered by the uneducated who (like in all religions of divine revelation) were ignorant, full of doubt and ungrateful.

In an attempt to account for Brahmgupta's sin against conscience, Alberuni has offered two conjectures. First, the Indian astronomer might have been struck by a calamitous fate like Socrates, notwithstanding the abundance of his knowledge and the sharpness of his intellect. The second possible explanation lay in his extreme youth, for he wrote the Brahmasidhanta when he was just thirty. In other words, he propounded his theories when he had not acquired enough experience and maturity.[11]

It is safe to surmise from Alberuni's observations that the educated Indians nurtured a dual attitude – scientific and non-scientific – towards the natural phenomena. Even after the passage of three centuries, this particular outlook was discerned in Amir Khusro (1253–1325) who was otherwise unsurpassed for his erudition.[12] When the famous Persian poet-cum-historian assumed the task of demonstrating the superiority of India over other countries, he built up a case that was a curious mixture of fact and fiction. He found no contradiction in fusing factual information with legendary tales. He would have us believe that India's climate was similar to that of paradise because Adam, after his banishment from heaven, found shelter in this country. During the course of his search for Eve, Adam did not ease himself in India, as such acts were not permitted in a place like paradise.[13] Though peacock was a bird of heaven, yet it was found in India. The snake, who had intruded into heaven by a stratagem, was also found in India. Since it possessed the un-Indian habit of biting, it was rendered harmless by being made to live beneath the earth. The climate of India was superior to that of lands in central and west Asia. Unlike Khurasan, its winter was not cruel and a single garment was sufficient to keep warm. Its summer did cause discomfort, but the shade of a single branch of some tree provided protection against the blazing sun. According to a *hadis*, the cool breeze felt in Hijaz blew from the direction of India. The spring, characterised by the availability of wine and flowers, lasted throughout the year. The flowers blossomed for a long time and retained their fragrance even after drying up. Juicy fruits included mango, banana and sugarcane. Dry fruits included cardamom, camphor and cloves. An accompaniment of food, betel leaf, was relished by the nobles and their sons. Even the poor people of India were not avaricious, being satisfied with a mere blanket. A Hindu peasant, wearing a single worn out sheet of cloth, spent the night in a pasture and grazed a single animal.[14]

Amir Khusro felt that the contemporary Muslim elite carried a prejudiced view regarding the religion of Hindus and the state of their sciences. He assumed the task of putting the record straight and clearing the misunderstandings. Having cultivated the friendship of Brahmins and having won their confidence, he undertook independent research into the secrets of their learning. He found that whatever he had grasped from the Brahmins could not be contradicted. He succeeded in identifying the factors that did not permit the Brahminical learning from being disseminated on a wide scale. Some branches of this knowledge were excessively complicated, so that the mediocre did not dare to pursue them. The Brahmins, owing to their humility, generally remained silent and did not speak eloquently about their learning. Some people (apparently orthodox Muslims) regarded the Hindus as non-believers and therefore did not acknowledge their virtues. Those who (like Amir Khusro) did not share this view, held that Hindus were pantheists and believed that an all-powerful God sustained the universe and that every action in this world emanated from His will. They knew that even while the Hindus worshipped several physical objects – idols, sun, stones, cattle and plants – they regarded these things to be the creation of God and perceived a divine spirit in them.[15]

Let us consider the perceptions of Amir Khusro on the state of Indian sciences and the contribution of Indian scientists. Since he was an 'insider' motivated by patriotic fervour, his observations formed a counterpoise to Alberuni's unfavourable judgements as an 'outsider'. In his opinion, India was superior to the whole world with regard to learning and mental illumination.[16] A huge wealth of knowledge had been created in such diverse disciplines as natural sciences, mathematics, geography, logic, astrology and poetry. The Brahmins had surpassed Aristotle and Rumi in their contribution to philosophy. Such was their perfection in various sciences that they did not deem it necessary to go to other lands to improve their knowledge or to seek a judgement on it. Instead the foreigners travelled to India from all parts of the world to acquire knowledge in different subjects. The science of mathematics, as developed by a Brahmin named Asa, was borrowed even by the Greeks. A foreigner Abu Maashar resided in Benares for ten years and acquired unprecedented proficiency in astrology. The wonderful book of wisdom, *Kalila o Dimna*, had been translated into the major languages of the world and had provided inspiration to scholars across the globe. The fascinating game of chess, with its countless moves and innumerable possibilities, did not permit anyone to grasp its technique. The Indian music was so complex that foreigners could learn it only after staying in India for thirty or forty years and so captivating that it could charm the beasts.[17]

Amir Khusro would have us believe that some Indians had acquired supernatural powers, by virtue of which they could perform miracles. Conscious of the improbability of the alleged feats, he admits that certain things were achieved through magic, while others were just popular stories. He observes, "One who has seen them cannot forget them and one who has not seen them, does not believe them". For example, a person bitten by a snake was revived even after six months by being flown eastwards to Kamrup (Assam) for being treated by magic. A dead man could be revived by the recitation of incantations (mantras) by Brahmins specialising in occult, provided the vital organs of the dead had not decomposed. Some jogis could achieve extraordinary long lives by practising breath control. Jogis inhabiting the mountainous caves of Kashmir could transfer the soul from one body to the other. They could assume the form of different animals, draw out the blood of a person and refill it again, control the minds of the young and old, fly in the air, remain under water without sinking, sit cross-legged on the surface of water and cross the river in that posture, become invisible by applying collyrium to the eyes, make the clouds shed or stop the rain, and so on.[18]

Amir Khusro has devoted a section of his description on India to a few birds and animals – peacock, parrot, skylark, monkey and elephant. Here was an opportunity for him to make scientific observations on living creatures, focusing on their normal behaviour in the natural habitat. However, he missed the opportunity, as he highlighted some unnatural characteristics imposed on them by their human trainers and which were magnified beyond proportion with a view to rouse the reader's sense of wonder. It was true that that these observations carried little significance from standpoint of biological sciences.[19] Yet it was safe to deduce from the descriptions that there existed a class of professionals who trained birds and animals for the amusement of upper classes. More significantly there were biologists who undertook research and experimentation into a wide range of animal behaviour including chirping of birds, running of deer and howling of jackals. They were reputed to have written a large number of books on the subject. Had these books been extant, they would have served as an index of the state of life sciences in medieval north India.

II

We notice a marked disjuncture between the thirteenth and fourteenth centuries, when we analyse attitude of the ruling class of the Delhi sultanate towards the process of state formation and the indigenous society. During the thirteenth century, the Ilbarite rulers employed reactionary devices to build a new administrative structure on the ruins of defunct feudal polities. Iltutmish and Balban traced their descent to Afrasiyab, the mythical Persian hero, and forbade the recruitment of low-born persons to official positions even in the lowest rungs of a nascent bureaucracy.[20] The sultans sought to establish the hegemony of narrow ethnic groups, while paying little heed to professional efficiency and administrative skills. The ruling class possessed little knowledge about the indigenous society, social realities and aspiration of diverse social groups. They knew little about the ecological features, patterns of production, dynamics of internal trade and economic basis of urban centres. They failed to institutionalise the collection of statistics and maintenance of records, owing to limited availability of paper and trained manpower. The emoluments of military commanders and their soldiers were never stated in precise figures. The land revenue was not assessed and collected from individual peasants or single villages. Instead, the ruling class extracted irregular tributes from the local chiefs and this too after undertaking frequent military expeditions in the unpacified countryside.[21]

From the early fourteenth century onwards, winds of change began to blow across the Delhi sultanate, as it entered a higher stage of state formation. The forces of orthodoxy were not entirely suppressed, but they were effectively marginalised. The state policy was not based on the canonical law, but on the needs of a pragmatic statecraft. The ethnic base of the bureaucracy was broadened, as the Turkish domination was terminated and the non-Turkish elements were given their due. Muhammad bin Tughluq assimilated foreigners, sufi families and socially depressed elements into the state services.[22] He suppressed the orthodox theologians and encouraged atheistic philosophers like Ubaid Shair, Saad Mantaqi, Najm Intishar and Maulana Alimuddin.[23] Though it was not possible to determine the impact of these measures in precise terms, yet it was safe to suggest that a new intellectual environment was created in which the traditional modes of thinking were undermined and creative solutions were applied to the problems of governance, particularly the role of state in restructuring the agrarian economy.

The centralisation and systematisation of administration, as pursued by the ruling class of the Delhi sultanate from the beginning of the fourteenth century onwards, was manifested in an emphasis on the collection of statistics and maintenance of records. The emoluments of nobles were determined on the basis of their personal rank and the size of their contingent, while their financial obligations were matched with the income from their revenue assignments (*iqtas*). Efforts were made to determine the land tax expected from different localities. The salary of a horseman was fixed at 234 tankas, while the soldier possessing an extra horse was entitled to an additional sum of 78 tankas. The ruling class familiarised itself with the relative position of various social classes in the rural areas. The role of intermediaries in the agrarian economy was comprehended, while they were deprived of their traditional privileges and converted into semi-official components of the revenue collection machinery. An enormous exercise was undertaken to collect information regarding the actual area under cultivation, yield per unit of area, the number of houses in a village and even the cattle in possession of the peasant. Much of this data appeared to have been recorded in the account books (*bahis*) of the village accountants (*patwari*). Since the land tax was increasingly collected in cash, the movement of agricultural prices became crucial. The entire system, as described above, necessitated the employment of a virtual army of functionaries – clerks, writers, accountants, assessors and agents – who were adept in Persian as well as the local language, besides possessing the knowledge the new revenue regulations and requisite skills in writing.[24] In consequence, large sections of the society from the mighty noble to the lowly peasant – central revenue department, military commanders, assignment holders, land grantees, revenue officials, grain merchants, moneylenders, intermediary chiefs, village headmen and of course the primary producers – were drawn into the vortex of mathematical calculations, revenue statistics and official records.

The ruling class of the Delhi sultanate was convinced by painful experience – droughts, famines, scarcity of food grains, exorbitant prices, starvation deaths and abandonment of villages – that agricultural production could no longer be left to the vagaries of nature. Building on its recent success in political centralisation, it decided to tame the natural forces and intervene in the agrarian economy, with a view to boost agricultural production and revenue resources. In response to the seven years long famine (1334–1341), Muhammad bin Tughluq created a department for the promotion of agriculture (*Diwan-i-Amir Kohi*), but the contemporary sources have failed to preserve any record of its discussions and thus deprived us of valuable insights into the perceptions and concerns of the medieval agronomists. However, under an elaborate project of agrarian expansion, liberal loans were given to designated officers (*shiqdars*) who promised to bring thousands of acres of waste land under the plough, along with a radical shift in the cropping pattern – wheat in place of barley, sugarcane in place of wheat, and grape and date in place of sugarcane.[25] In the wake of the failure of this scheme, the ruling class shifted its focus of agrarian expansion from the Ganga-Jamuna Doab to the Sutlej-Jamuna Divide. During the reign of Firoz Shah Tughluq (1351–1389) a network of canals – which took off from the Sutlej, Ghaggar and Jamuna – provided artificial means of irrigation to the dry tracts of Haryana, so that the spring (*rabi*) crop began to be raised in Hissar, where the rain grown autumn (*kharif*) had been the rule.[26] Though Barani did not live long enough to observe the resulting transformation in the countryside, yet his speculation on the subject was quite rational. The following passage shows that even a diehard conservative in socio-political matters like Barani was capable of doing scientific thinking on restructuring of the agrarian economy.

Barani anticipated fundamental and widespread changes in the rural areas of the Sutlej-Jamuna Divide, owing to the network of canals that were excavated in the middle of the fourteenth century. Groups of pastoralists, who moved from place to place in search of water along with their families and herds, would settle as sedentary cultivators. Owing to the availability of water, superior crops like wheat, gram and sugarcane would be raised in places where pulses, lentils and sesame were produced. The process of colonisation would be set in motion, so that a large number of new villages would be settled and the revenue assignees (*muqtis*) would be able to collect the land tax regularly from them. Large armies would be able to encamp for decades on the banks of these canals, owing to the easy availability of water. Due to general prosperity and improved living standards, people would be able to consume superior foods – wheat, gram and sugar – which earlier were enjoyed only on festive occasions and were brought only from Delhi and its neighbourhood. Since a surplus of these commodities would be produced, they would be transported to deficit areas for commercial purposes. People would be able to travel across the waterless tract, without carrying any stock of drinking water. They would also be able to travel in boats through the new canals that were fifty or sixty *kos* in length. A desert like area, where nothing grew except acacia and thorny bushes, would be covered with gardens where a variety of fruit and flowers would grow. The birds and animals would no longer suffer due to the scarcity of water.[27]

III

The bhakti reformers occupied a considerable space in the socio-cultural life of northern India during the pre-Mughal times.

Here we are concerned with two aspects of their role viz., the content of their social-religious ideologies and the manner of their articulation and dissemination. While pleading for devotional monotheism and social equality, they sought to demolish the hegemony of rulers and priests. Their message was specifically meant for the socially depressed groups – peasants, artisans, labourers and women – who suffered from economic exploitation and denial of access to education. They sought to rescue the under-privileged from the stranglehold of ignorance and superstition, which had been cleverly imposed by the pandit-mulla regime. Assuming the role of stern teachers and loving mentors, they communicated with their followers in the common dialect and avoided high-flown philosophical jargon. They found that the masses had been fed on myths and miracles regarding the creation of universe and the evolution of life on earth. The bhakti reformers explained the laws of nature in scientific terms and illustrated their arguments with examples from the ordinary experiences of the common masses, who could again verify them by simple observation. In this manner the bhakti reformers – particularly Kabir and Guru Nanak – induced the oppressed and illiterate, both directly and indirectly, to adopt a scientific attitude towards their immediate surroundings as well as the wider universe.[28]

Kabir was an acute observer of nature.[29] His observations on the natural phenomena were based on scientific principles. He argued that the characteristics and behaviour of living creatures – plants and animals – was governed by the laws of nature and that these could not be altered by any external agency through the application of force. Nature was supreme in its domain and did not brook any interference. It was incumbent on human beings to form a correct understanding of natural phenomena, particularly those that touch the human life. The truth regarding the laws of nature was self-evident. For example, the cloves did not bear fruit, the sandalwood tree did not have flowers, fish did not hunt in the forest, the lions did not swivel in the sea, castor oil plant could not turn into Mysore sandal, a blind man could not see, a cripple could not leap over the Mount Sumeru, a dumb person could not express his knowledge in speech, there could be no tree without roots, plants could not bear fruit without flowers, nobody could dance without feet, praise could not be sung without the use of tongue, and so on.[30] In a brilliant contrast, Kabir argued that the social status of an individual could change due to worldly factors, but the same could not be said regarding matters that were in the hands of nature.

In a series of razor-sharp arguments, Kabir has demolished the claim of Brahmins that their lives remained untouched by flesh. The

Brahmin's house was made of clay in which the dead bodies of human beings and animals had been decomposing for thousands of years. He drank the water of the Ganga and regarded it as pure, but it was contaminated with the flesh and blood of thousands of creatures – fish, turtles and crocodiles – that had been breeding on its banks. He consumed milk everyday after lunch, but it was produced in the organs of milch cattle that were made of flesh and blood. Therefore he was advised to throw away his scriptures, which had failed to take note of the most glaring realities of life.

Kabir was strongly opposed to animal sacrifice as an act of religious merit. He condemned both the pandit and mulla for this widespread cruelty. How could the Brahmins claim that the animals were slaughtered to please the gods? Had it been so, the gods themselves would have gone to the jungles to hunt the beasts. Instead of reaching the gods, the meat ended up in human bellies. The animals were created by Brahma, the potter, in the manner he shaped other forms of creation. But man was engaged in the mass killing of animals, as if they grew season after season like grass. The Muslims too slaughtered cattle as a part of their religious rituals. They converted a living creature into a corpse and still claimed that it was a holy rite. They asserted that the practice was in conformity with the preaching of their forefathers and that it had been sanctified by long tradition. But the blood, which had been shed since then, was on their head. They must share this huge backlog of guilt. They did not know that the animals were born through the same process of sexual reproduction (merger of blood and sperm) as was found in all other creatures. Despite an entire life spent in continuous prayers, they had failed to comprehend the true basis of religion. The association between animal killing and religious merit was untenable, as the real purpose was to satisfy the hunger for taste. Both the pandit and mulla, owing to their cruelty towards animals, would go straight to hell.[31]

Since Kabir believed in the universal brotherhood of man, he rejected divisions on the basis of religion and caste. All human beings passed through the same life cycle from birth to death. Nobody read the Vedas in the womb and no Turk was born circumcised. Every human embryo grew from a single seed, while every foetus grew in the mother's womb for ten months. When the babies were born, they were composed of the same material viz., skin and bone, flesh and blood, piss and shit.[32] During their active life, all human beings showed the same traits viz., hunger, anger and lust. When they died, their corpses ended up in the same fashion, irrespective of the mode of final disposal. If the dead man was cremated, his burnt ashes mingled with the dust. If he was buried, his dead body was eaten by the maggots. If it was neither cremated nor buried, but it was left unattended, then it became the food for pigs, crows and dogs.[33] These were hard facts of life, from which there was no escape. Since death was the supreme and universal reality, it did not recognise any social divisions.

Guru Nanak (1469–1539) came out strongly against the priestly classes for misguiding the common people regarding food habits, eating of meat and vegetarianism. In the Brahmanical religious discourse, references to flesh occupied a crucial place. The Brahmins propagated that abstaining from flesh was an essential component of religious practice and that it led to spiritual betterment. Guru Nanak found that such beliefs were not only opposed to human experience, but also came into conflict with the laws of nature. He felt that such beliefs generated a plethora of superstition, contributed to mass ignorance and amounted to intellectual oppression. Therefore, he entered into a direct confrontation with the priestly classes on the issue in the following manner.[34]

The Brahmins regarded themselves as wise in religious and spiritual matters, but they knew nothing of meditation and gnosis. They wrongly equated the distinction between virtue and vice with that between vegetables and meat. The ethical qualities of human beings could not be determined by their preference for particular types of food. The Puranas allowed the eating of flesh, the semitic books permitted it and it was consumed during the four ages. It was customary for gods to kill the rhinoceroses and burn the libations of their flesh in the sacrificial fires. Flesh had been an esteemed food in their yajnas, marriage feasts and religious rites. But now the Brahmins abjured flesh and sat holding their noses in the semblance of meditation. They were born of the blood of their parents, but refused to eat flesh or fish. Only fools wrangled over the eating of meat. Human beings could not escape from their association with flesh.[35] Sex was a basic human need, but copulation was nothing but a meeting of the flesh of two persons, male and female. When a man and woman met at night and engaged in sexual intercourse, conception took place in the flesh. All human beings – men and women, kings and emperors – were not only born of flesh, they were also composed of flesh. If association with flesh led the concerned person to hell, why did the Brahmins accept offerings from him. It was ironical that the giver of charity went to hell, while recipient of charity went to heaven. They were absolutely ignorant of the manner in which flesh was produced, because the process was not dissimilar to the cultivation of crops – grain, sugarcane and cotton – which were raised with the help of water.

Guru Nanak was anguished at the prevalence of widespread misconceptions regarding the common human experiences viz., birth, growth and death. The contemporaries also nurtured strange notions regarding the structure of female body, the inner biological changes, the monthly menstrual cycle, the conception and growth of foetus in the mother's womb, the birth and early nourishment of the new born, and so on. On the one hand, the currency of these notions negated the laws of nature and, on the other, contributed to the degradation of women in the society.[36] For example, during her menstrual cycle, a woman was regarded as polluted and therefore avoided. The process of menstrual bleeding was regarded as shameful and degrading, being equated with sickness. Armed with a rational understanding of the laws of nature, Guru Nanak revolted against the above attitude and reprimanded those who stigmatised as polluted the garment stained with female blood.[37] He held that the father and mother were equal participants in the sexual union. The merger of mother's blood with the father's semen resulted in the formation of an embryo. The foetus grew in the mother's womb, getting warmth and protection for nine months. Soon after taking birth, the baby was attracted to the mother's breast from which it drew its sustenance and nourishment.[38]

In conclusion, we would consolidate the different parts of the above discussion. We have seen that Alberuni, while appreciating the Indian contribution to the advancement of astronomy and mathematics, identified the factors that inhibited the growth of scientific knowledge and noted a perpetual streak of duality – a mixture of scientific truths and mythical beliefs – that characterised the research of leading scientists. Amir Khusro's passionate assertions regarding the superiority of Indian sciences assumed, to an extent, the form of a counterpoise to Alberuni's negative judgements. But these observations were marred by patriotic fervour, apologetic tone and penchant for highlighting the wondrous and supernatural, rather than the natural and ordinary. From the beginning of the fourteenth century, circumstances were created for the application of scientific ideas to the problems of governance and economy. The centralisation and systematisation of administration, as pursued in the Delhi sultanate, produced some unintended consequences. An emphasis on collection of statistics and maintenance of records, coupled with the increasing availability of paper, induced a large number of people – ranging from the high ranking noble to the lowly placed peasant – to accept a new way of life, where quantities were expressed in terms of precise figures. A more effective social role appears to have been played by the bhakti reformers who undertook the more fundamental task of fighting ignorance and superstition, which had been perpetuated by the priests and theologians. By assimilating the laws of nature into the socio-religious discourse, they persuaded the lower social orders to adopt a scientific attitude towards their immediate surroundings as well as the wider universe.[39]

Notes

1. Alberuni (AD 973–1049) was born in Khwarizm during the Samanid rule. His life was affected by violent changes in Central Asia. From 1017, he lived in Ghazni as a virtual hostage of Sultan Mahmud. He appears to have visited Punjab and other parts of northern India. He was proficient in such languages as Arabic, Persian, Hebrew, Syriac and Sanskrit. A prolific writer, he is reputed to have produced over 114 works on diverse subjects. His observations on India embrace religion, philosophy, law, science, metrology, weights and measures, alchemy, geography, cosmography, astrology, astronomy, chronology, social structure, customs and manners.
2. Edward C. Sachau, *Alberuni's India*. Vol. I, New Delhi, reprint, 1964, pp. 17–19 (hereafter cited as Alberuni).
3. For comparison, consider an analysis of Indian society on the eve of Turkish conquest, with reference to cosmology and sciences, as observed by Alberuni K.A. Nizami (ed.), *Politics and Society during the Early Medieval Period: Collected Works of Professor Mohammad Habib*. Vol. I, New Delhi: Peoples Publishing House, 1974, pp. 180–84.
4. Alberuni, I, op. cit., pp. 22–24.
5. Ibid., p. 25.
6. Ibid., pp. 264–65.
7. Varahamihira included a growing interest in horoscopy and astrology to the study of astronomy and mathematics, an attempt that went against the theories of Aryabhatta. Romila Thapar, *The Penguin History of Early Medieval India: From the Origins to AD 1300*. New Delhi: Penguin Books, 2002, p. 307.
8. Alberuni, II, op. cit., pp. 108–10.
9. The dual attitude of Brahmagupta could be understood by moving forward one thousand years and considering the contradictory position of Rene Descartes, the European scientist. Debiprasad Chattopadhyaya, *History of Science and Technology in Ancient India*. Vol. III, Calcutta: Firma K.L. Mukhopadhyay, 1996, pp. 75–79.
10. Alberuni, II, op. cit., p. 113.
11. Ibid., p. 112.

12. For studies on the life and career of Amir Khusro, see Mohammad Habib, "Hazrat Amir Khusrau of Delhi", in K.A. Nizami (ed.), op. cit., pp. 291–355; Wahid Mirza, *The Life and Works of Amir Khusrau*. New Delhi, reprint, 1974, pp. 16–139; K.A. Nizami, *Studies in Medieval Indian History and Culture*. Allahabad: Central Book Depot, 1966, pp. 102–21.
13. Amir Khusrau, *Nuh Sipihr*. (Persian text), Wahid Mirza (ed.), Calcutta, 1950, 152–56; for an English translation of these passages, see R. Nath and Faiyaz Gwaliari, *India as Seen by Amir Khusrau*. Jaipur: Historical Research Documentation Programme, 1981, pp. 36–39.
14. Ibid., pp. 158–61.
15. Ibid., pp. 162–65.
16. Ibid., pp. 166–71.
17. For a recent analysis of these assertions regarding the superiority of India's climate and sciences, see Syed Ali Nadeem Rezavi, "The Idea of India in Amir Khusrau", in Irfan Habib (ed.), *India: Studies in the History of an Idea*. New Delhi: Oxford, 2005, pp. 122–27.
18. Amir Khusrau, *Nuh Sipihr*. op. cit., pp. 192–94.
19. Ibid., pp. 181–91.
20. Influence of Sassanid monarchical traditions and a deep-rooted prejudice against the low-born constituted the core of the political ideology of the Delhi sultans. Khaliq Ahmad Nizami, *Religion and Politics in India during the Thirteenth Century*. New Delhi: Munshiram Manoharlal, reprint 2002, pp. 95–117.
21. Tapan Raychaudhuri and Irfan Habib (eds.), *The Cambridge Economic History of India, Vol. I, c. 1200–c. 1750*. Hyderabad: Government of A.P. Press, reprint 2004, p. 60; on his return from the expedition against Malik Chajju, Sultan Jalaluddin Firoz Khalji chastised numerous local chieftains and extracted tributes from them. Amir Khusrau, *Miftah-ul-Futuh*. (Persian text), Shaikh Abdur Rashid (ed.), Aligarh, 1954, pp. 22–23.
22. Mohammad Habib and Khaliq Ahmad Nizami (eds.), *A Comprehensive History of India, Vol. V: The Delhi Sultanate, AD 1206–1526*. New Delhi: Cambridge University Press, reprint 1982, pp. 561–65.
23. Ziauddin Barani, *Tarikh-i-Firozshahi*. (Persian text), Calcutta, 1860–62, p. 465. (hereafter cited as Barani).
24. Barani, pp. 287–88; Yahya Ahmad Sirhindi, *Tarikh-i-Mubarakshahi*. (Persian text), Calcutta, 1931, pp. 101–02.
25. Barani, pp. 498–99.
26. Shams-i-Siraj Afif, *Tarikh-i-Firozshahi*. (Persian text), Calcutta, 1890, pp. 127–28.
27. Barani, pp. 567–70.
28. The contribution of bhakti reformers appears to have stood in complete contrast to that of the sufis who, it has been alleged, opposed science and philosophy. Irfan Habib, "Reason and Science in Medieval India", in

D.N. Jha (ed.), *Society and Ideology in India: Essays in Honour of Professor R.S. Sharma*. New Delhi: Munshiram Manoharlal, 1996, p. 164.

29. There is much scholarly debate on almost every aspect of Kabir's life viz., date and place of birth, social background, religious affiliation, source of inspiration, interpretation of verses etc. Charlotte Vaudeville, *A Weaver Named Kabir*. New Delhi: Oxford University Press, 1993, pp. 39–65.
30. Linda Hess and Sukhdev Singh, *The Bijak of Kabir*. New Delhi: Motilal Banarsidass, 1986, p. 49.
31. Ibid., pp. 57–58.
32. Ibid., p. 69.
33. Ibid., p. 67.
34. Ibid., p. 65.
35. For an analysis of Guru Nanak's teachings on the basis of his devotional verses and a comparison with the socio-religious thinking of other bhakti reformers, see W.H. McLeod, *Guru Nanak and the Sikh Religion*. New Delhi: Oxford University Press, 1968, pp. 148–226.
36. Bhai Jodh Singh, *Gospel of Guru Nanak in His Own Words*. Patiala: Punjab University, 1969, pp. 138–39.
37. Nikky Guninder Kaur Singh, *The Feminine Principle in the Sikh Vision of the Transcendent*. Cambridge: Cambridge University Press, 1993, p. 55.
38. *Adi Granth*. p. 140.
39. Ibid., pp. 75–137.

16

Mining and Metallurgy in Medieval Deccan

S. Jay Kishan

The geographical and cultural characteristics of Deccan have a unique and distinct place in the subcontinent. It has achieved distinction in the metallurgy and industry due to its topographical features since ancient times. Medieval period has witnessed considerable development in manufacturing of arms due to various factors. Induction of artillery was a significant contributing factor. Artillery has accelerated the mining of ores, processing of sulphur and saltpetre. Industry of arms attained distinction in this period. In these sectors several thousand people were employed, in various activities related to mining, processing and manufacturing of arms, to cater to the demands from the locals buyers and external traders. The scientific techniques of production of arms, mining of ores, processing metals were not fully developed. Yet, this period was characterised by full development. It was characterised by the application of extraordinary craftsmanship while manufacturing various articles, arms, armour and artillery pieces.

The object of this paper is to focus on the various mining and manufacturing centres in Deccan and also an attempt has been made in this paper to focus on adopted technology in the process of iron and steel industry, particularly processing of ores, metal processing, foundry, and smithy. I have based my paper on contemporary Telugu texts, Kaifiyaths, translated Persian records, Travelogues and monumental secondary sources.

Since ancient times India was famous for good iron resource. "... India was considered as the centre for establishing iron and steel industry in the world. It is probable that the Iron Age in India started about 3000 years earlier than the west. That time Hyderabad and Madras in South India were the centres of production of Wootz."[1] It was the cheapest metal available in abundance. The iron and steel industry was a vital industry in India.

Deccan armies in the medieval period have equipped themselves with huge number of arms and good quantities of ammunition in the forts. War institution was the greatest employment generating area in the medieval period. Thousands of men were involved in mining, processing and making of war equipments. Equally good number of men were recruited in the military service. They were provided with arms, armour and ammunition, either by the Kings or 'Jagirdars'. All the forts in Deccan either big or small were equipped with good number of cannons. They were kept ready for move into the warfields and also entrenched on the pivot projecting outside through the embrasures. Matchlock guns were provided to the soldiers and reserves of ammunition were maintained. In the middle of the 14th century Bahamany and Vijayanagara kingdoms strengthened their military force for the hegemony in the south, This lead to strong conflicts and many wars between these two kingdoms. Frequent wars between Bahamany and Vijayanagara necessitated the increase in production of arms and ammunition.

Different kinds of metals were mined and processed by the smelters in their crucibles. Different kinds of articles were manufactured in local 'Kharkhanas'. Iron, copper, lead, bronze, steel metals were used in manufacturing of arms. Sulphur, saltpetre, wax and lacquer were procured and was processed in huge quantities for the purpose of wars. The mining of ores, processing and manufacturing of metals were done in various places in Deccan. Smiths, workers and labourers were very briskly engaged. Smiths and workers, not only native people, but also foreigners were engaged in smelting, foundry and smithy.

Weapons industry flourished extensively in this period. The industry attracted natives as well as people from other places. Foreigners were involved in manufacturing of weapons, cannons and matchlock guns in particular. There was enhancement in the production of cannons, matchlock guns and cannon balls.

Europeans and Central Africans, who were known for their specialisation and expertise in this technology, did the manufacturing of artillery pieces in the local foundries. "In the middle of 16th century good number of workers' communities migrated to Deccan."[2] They were employed and granted revenue free lands. Particularly Europeans and Central Africans were employed in the foundries of artillery and Turks and Persians were employed in arms manufacturing 'Kharkhanas'.

In the medieval period, iron industry attained prominence. The application of art and science to this metal (iron) by the native artisans for extraction of ores, refinement, alloying, shaping into various articles and applying extraordinary craftsmanship was in vogue in the medieval Deccan. The availability of good quality of iron ore in abundance in superficial layers of crust of earth and its easy extraction were main reasons for the development of the iron and steel industry in the Deccan. Iron ore deposits were richly distributed in the Deccan and this factor facilitated mining and establishment of manufacturing centres.[3] War requirements necessitated the establishment of good number of 'Kharkhanas'. Those Kharkhanas producing artillery equipments were known as 'Madrsa-i-Sanae'. The application of scientific techniques to develop this metal was unknown to the artisans in this country. There was good quantity of iron, and steel production was done at various places in this region. There was abundance of good quality of iron ore mines in Golkonda, Bijapur and Ahmednagar regions in the Deccan.

The iron ore was extracted and smelted at Warangal, Konasamudram, Dindurty, Kamarapally, Errapally, Mulkanur, Nirmal, Gudikol, Mylavaram, Ravata, Kalyani and many other places.[4] Nirmal in Adilabad district and Indur is present Nizamabad in Andhra Pradesh state were known for rich resources of iron. "Nirmal and Indur possessed iron ore mines, arms production center and other metals."[5] "The best steel was produced at 'Konasamudram' near Nirmal in Nanded district."[6] Thevenot on his way to Golconda visited the above places and he described "... the town Indalwai of which nothing is to be said in particular but that great many swords, daggers and lances are made there, which are vended all over the Indies and that iron is taken out of a mine near the town, in the mountain of 'Calagatchi' (Kalaghat)."[7] Hyderabad iron was famous for its best quality (It was in 'swatha varna' white in colour) "... since it was used as raw material for Damasin blades."[8] The best variety of steel ingots was produced at Konasamudram near Nirmal. This iron was widely used in the Nirmal industry of arms in the medieval period. Dindurthy mines produced the best variety of steel marketed for Damascus. "The mines here being mere holes through the thin grain tic soil, from which the one is detached by means of small iron crowbars."[9] Iron ore deposits were abundantly available at superficial layers of the earth crust. It was an easy task for the miners to locate the deposits of ore, and to dig out with the help of labour in small quantities and smelted in their local crucibles, turning the metal into small ingots. Daniel Havart a Dutch traveller visited Golkonda between AD 1672–1685. He visited several places in the region and described about mining of iron and smelting centres.[10] He stated "... Abundance of good quality of iron ore at Golkonda region on the surface level was one of the reasons for the presence of good number of mines and manufacturers."[11] "During 16th and 17th centuries mining and manufacturing work was very brisk. "Iron mines were located at Indur, Nirmal, Nagalwancha, Machilipatnam, Ramayapatnam, Bhimunipatnam, Mootapalli and Singarayakonda."[12] Havert mentioned about these places. But there were several other places of iron manufacturing centres including Warangal, Konasamudram, Dindurthy, Kamarapally, Errapally, Mulknoor, Gudikole, Mailavaram, Ravancha, Jagitial, Yelichala, Ramayampet, Konapuram, Kallur, Ananthagiri and Lingampally in the Golkonda region.[13] Havert further stated "The iron stone was located commonly at a depth of about a persons height at the foot of hills. Simple tools like crowbars and handspikes were used for digging of the ore. Innumerable deposits of Iron ore and varying qualities were widely distributed all over the lateritic tracts of Deccan."[14] Musali Madugu and Dachepally (in Cuddapha district) were other places in Golkonda region where excellent quality of steel was produced.[15] At Dachepalle the iron ore was brought from the nearest Nallamalli Hill range which was about five miles to the east of the village. "The ore was obtained by quarrying the hill 12 feet deep."[16] Quarrying the ore was not taxed by the state. There was a practice of leasing mines. The practice was seen in regarding diamond mines. Qutb Shahis and Adil Shahis were collecting good revenue from the diamond mines. They were having good command over the diamond miners but not on the iron ore quarries.[17] Miners were free to sell the ore anywhere in the local markets by carrying in the baskets. Sometimes tribes and locals were quarrying the ore. They were selling in the nearest village markets. Village smiths used to buy the ore from these local miners or from the itinerant traders. The tax was levied on the traders, artisans and merchants on their shops, Kharkhanas, stalls and instruments by the person called 'Mutharif'.[18]

In the kingdom of Bijapur, iron ore deposits were found in several places. Mining and manufacturing of iron and steel was very brisk. "The rocks of Kangtee (Kangtee in Asifabad, in Adilabad district) which are extensively developed between the river Godavari and Wardha valleys abound in hard ferruginous pebbles and clay iron ores."[19] Ahmednagar was an important cannon casting centre in medieval period. The world's most popular ordinance piece 'Malik-i-maidan' (King of the battlefield) which is now on the ramparts of Bijapur fort. Another cannon 'Kark bijili', one that is of the same size as Malik-i-maidan, was sunk in the river Krishna after the battle of Tallikota. Both the cannons were moulded during the reign of Burhan Nizam Shah in 1548. The technical work was done under the supervision of a Turk named Mohmmed, son of Hussain Rumi. [20]

'Malik-i-maidan' the biggest incendiary piece was a testimony to the excellent craftsmanship of artisans of Deccan in India.

Inscriptional evidence found on large number of cannons, which exist in various forts and housed in various museums, reveal the names of the cannons and makers of those cannons. It also reveals the names of those officers who controlled them. Iron was not only mined but also found on surface layers of earth. The lumps of meteorite iron was collected and those lumps were turned into the best steel. [21] Aurangabad was a popular centre for iron market in the medieval period.

Primarily there were two stages in processing of the ore. First collected iron ore was washed with water and smelted in the local crucibles. This iron ore was smelted in crude country furnaces and moulded in a malleable state and cut into small pieces weighing about a one-fourth share or a half share. These impure iron ingots were further smelted and turned into fine steel at famous steel production centres in the Deccan. However, cumbersome and obsolete techniques of production were in practice. This crude method continued till the beginning of the 19th century. In this process large quantity of iron could not be obtained at a time, but small quantities of iron ingots were produced by the same method. Therefore, iron was not easily converted into steel in large quantities.

The iron ore was separated from quartzite mixture by washing. Iron sand and was extracted and smelted in furnace for pig iron. The pig iron was further processed into cast iron and wrought iron. This was, comparatively an impure form of iron obtained by smelting pig iron. The ore was smelted with firewood without any flux and was obtained quickly in a tough and malleable state superior to any English iron or even the best 'Swedish'.[22]

This was used in moulding of guns and cannons. Wrought iron was a pure form of iron as it was produced from pig iron, and contained no carbon. Preparations of any kind of iron and steel depended largely on the chemical composition of the alloy. Heating and shaping could change its physical properties. Locally made country steel was known as *Kheri*. The finally processed steel or Persian steel was used for sword blades, and other sharper weapons were known as 'Foulad'.[23] The pure iron metal that turned into steel was grouped into three:

1. Carbon steel.
2. Alloy of steel.
3. Stainless steel or tools steel.

In this period the steel used in making arms was identified by its shine and luster coloured with shades of Caster oilseeds, and was known by different names such as 'Krishnapanjara' (Assam iron) 'Krishnasara' (Gujarath iron), 'Swetha Sara' (Maharashtra iron) ' Medikavarna' (Karnataka iron) 'Swetha varan' (Golkonda iron). 'Panjara', 'Swarna Panjara' and 'Nirasara', were identified as places of production.[24]

Havert describes the process of conversion of iron ore into metal and steel in the Golkonda region: "Iron ore was roasted in the *Bhatti* under the layers of fire wood after over heating the ore liquid metal plowed down into a cavity in the ground was quenched in water. The lump iron converted into bars, thereafter refining into steel. Steel was manufactured through various processes. The bar iron lumps were pulverised into pieces. These were put into earthen crucibles, fixed in the furnace and baked for 24 hours with charcoal then fire was allowed to cool down. At this stage small lumps of steel were found in each crucible."[25]

Dachepally Kaifiyath describes the process of manufacturing of iron. "The ore was brought from … was usually in the shape of granules … . Then they were ground in to powder and *puti* roused by one of the four workers employed. Normally a hearth was made, and on it a chimney of 3 feet high was constructed, while the bellows were fixed to one side of the hearth and a *Natakam* (a measure equal to four seers) full of iron ore was put in to the hearth with a basket full of charcoal to cover it."

"Two workers were engaged to blow the bellows, while the bellows were blown, another worker *Pannudugugadu,* prevented the flames from coming out of the hearth by closing it with wet mud taken out of *cheravala* which he held all the time in his hand, while another worker stirred the burning charcoal. After the charcoal settled down in the hearth another *Natakam* of iron ore was put into the hearth and process was repeated. This process was repeated for six times. When the smelted iron ore gathered at the pit of the earth in a liquid form, the hearth was opened and with the help of a spade, the smelted ore was taken out and allowed to cool. After the smelted lump cooled down sufficiently and got slightly solidified, it was lifted out with the help of tongs and it was put on wooden *tattees* where it was hammered out into rectangular block and cut into bars with an axe, each bar weighing twenty *seers*."[26] The process described in this Kaifiath seems to be little advanced method. But just before this period the primary smelting of ore, making into pig iron ingots or the Wootz, was prepared by burning in the firewood, which was abundantly available in this region.

The smelting centres were known as *Kamarasalas* or *Kotasalas* and run by traditional Hindu smiths. The native Muslims were also involved in the smelting, refining of iron and steel. They were involved in arms production and were considered as best manufacturers of arms and armour. The smiths and smelters used to purchase ore from minors or itinerant merchants or from the established markets in the towns known as *Aurangs*, where *'Beparies'* went to sell the ore or pig iron ingots. Again these iron ingots were made into fine steel.

The exhibits found in the Deccan museums are the works of famous sword smiths of this period. Asadallah Isafani of Persia was regarded as renowned sword maker in the world. The son and successors of Asadallah Isafani and his followers were associated with manufacturing of swords. The inscriptions on the swords revealed few names of the sword makers – Muhammed Misri, Mohamad Yusuf Misri and Muhammud-Ali-Misri were among the noted swords smiths and their names were inscribed on the swords. Mahammud-Ali-Misri was the master of sword makers in this period and was known as "Allama" (Master) Misri.[27] Thirumala Chidamabara, Chiligi Laxma Ramudu, Shanki Sumalu, Ashakusai, Subhabhaktha and father of the author of Khadga Lakshana Shiromani, Nalla Pichaiah were few skilled sword makers of this period mentioned in the 'Khadga Lakhshana Shiromani'.[28] 'The king Vijaya Raghunatha Thondaman of Puddukkota in Tamilnadu, honoured Nalla Pichaiah.'[29]

The smelting centres run by Muslims were called as 'Bhatti', situated in the towns. Small-scale smelting and refining of iron by village smiths was a regular practice in the villages. They were known as 'Kolimi', who catered to the domestic and agricultural needs in the villages. Iron blooms and cakes (Pig iron ingots) were sold in the markets. The different prices were mentioned for these articles in the contemporary records. The selling price of 'Hirakassi' (sulphate of iron) was 2.5 *sheers* for a rupee in Golkonda region in AD 1661.[30] The same selling prices prevailed in the Bijapur market.[31]

Surplus iron was produced in medieval Deccan and this facilitated export to other countries and the establishment of cannon foundries, which enabled casting of heavy cannons in Deccan foundries. Very long and big size cannons were built; weighing about 10 to 55 tons were caste in Deccan foundries. Early iron cannons were inaccurately built and consisted of wrought iron bars held together by rings and welded into a shape of cylinder. The use of such cannons caused failure in expeditions. To surmount this problem they produced single cast moulding, which was considered as an innovation made by Deccanis. It is quite evident that it was an improvement or an addition to existing technology due to the influence of the Europeans and Turks.

The technological changes were improved upon under the impact of Europeans. "In the middle of the 17th century, Dutch East India Company was organising the nails production in European style. The workers employed by them were only Indians who were very skilled in making nails as required by Dutch standards. In AD 1685. The Dutch used to purchase iron from indigenous smelters."[32]

"The Dutch East India Company was producing nails in Deccan and was exporting till AD 1886, from 100 kg's of iron 75 kg's nails were produced."[33] Sword blades and other weapons of war were extensively made at Hyderabad, Vanaparti, Gadwal, Kondapur, Jatprole and other places. As mentioned earlier those arms were not of good quality as compared to the imported ones. The imported weapons were of exorbitant rate and also were in demand. The imported sword blades from Damascus were fitted with local made hilts by the native artisans. Skilled labour was abundantly employed in iron mines, smithy and in making of articles and its ornamentation works. Babur exclaimed "Enormous supply of skilled labour in India."[34]

"'Indalwai' was the center for manufacture of swords, daggers and lances which were found in the markets all over the world".[35] It reveals that there was a significant development in iron and steel production. The cannons are the existing testimony for the craftsmanship of artisans. Several factors have contributed for this advancement. "Innumerable deposits of iron ore of varying qualities were widely distributed between Krishna and Thungabhadera rivers".[36] Mining, processing of ore and making of war equipments was at brisk. It attracted several traders from foreign countries. Abundance of natural resources attracted the European companies to establish their own manufacturing centres.

Notes

1. M.N. Paul, 'Industrial Development of India', p. 32.
2. Surjith Sen, 'Aspects of Indian Culture and Society', p. 140.

3. S.H. Bilgrami and C. Willmott (Compiled), 'Historical and Descriptive Sketches of His Highness, the Nizam Dominians', Vol. I, p. 140.
4. M.L. Nigam, *Catalogue of Arms and Armour in the State Museum.* Hyderabad: Govt. of A.P., 1975, p. 13.
5. *Ain-i Akbari*. Vol. II, Blochman English Translation, New Delhi: Oriental Books Reprint Corporation, 1977, p. 230.
6. S.N. Sen (ed.), *Indian Travels of Thevenot*. English Translation, New Delhi: The National Archives of India, 1949, p. 112.

   (Indur and Adilabad were in Nanded Sarkar. Present Indur is the district headquarters, Nizamabad and Adilabad also district headquarters, are in Andhra Pradesh State and Nanded in Maharashtra State)
7. Ibid., p. 112.
8. G. Bridwood, 'Industrial Arts of India', 1880, p. 170.
9. Ishranth Alam, 'Iron Manufacturing in Golkonda in 17th Century', Translation of Original Dutch Records of Danial Harvest. Paper presented in Indian History Congress 1988, p. 246. (Further it will be mentioned as Golkonda Iron by Ishrath)
10. Ibid., p. 246.
11. Ibid.
12. Ibid., 246. John Briggs, 'Tarikh-i-Qutb Shahi', English Translation, Appended in Vol. III, 'Rise of Mohammedan Power in India', Calcutta reprint, 1966, p. 272.
13. *Imperial Gazetteers of India*. Vol. XIII, 1958, p. 262.
14. George Watts and Percy Brown, *Arts and Crafts of India*. New Delhi: Cosmo Publication, 1979, p. 521.
15. G. Bridwood, op. cit., p. 170. Dachapalle Kaifiyath – Mackenzie Manuscripts, Document No. 311. Prof. Sorojani Regany, 'Pre-Industrial Rural Economy in the Ceded Districts as Reflected in the Mackenzie Manuscripts: A Case Study of Iron Industry at Dachepelle, Duvvur Taluq in the Cuddapah District'. The Department of History, Osmania University, Hyderabad, DRS Seminar Proceedings, 1995, pp. 18–26. (Further it will be mentioned as Dachepalle Kaifiyath)
16. 'Dachepalle Kaifiyath', p. 19.
17. Moughal Documents, Vol. I, Part II, AP State Archive, Hyderabad, 1658–63, p. 120.
18. 'Guntur District Kaifiyaths', Vol. I, AP State Achieves Department, Hyderabad, 1984, p. 18.
19. Dr S. Jaikishan, 'Artillery in Medieval Deccan', in *Studies in Indian History and Culture* (Commemorative Volume of Prof. Sarojani Regani), The Dept. of History: OU Hyderabad, 2004, 102–03.
20. Golkonda Iron by Ishrath, p. 24.

21. 'Khadga Lakshana Shiromani', by Navanappa, a Telugu poetic text of 18th century, described eight varieties and one hundred and forty kinds of swords in the book. The 1st variety of Patta swords were made up of this kind of iron, p. 4.
22. 'Golkonda Iron' by Ishrath, p. 24.
23. George Watts and Percy Brown, op. cit., p. 521.
24. Khdgalakshanamu, op. cit., p. 24
25. 'Golkonda Iron' by Ishrath, p. 227
26. 'Dachepalle Kaifiyath', pp. 19–20.
27. M.L. Nigam, op. cit., p. 19.
28. Khdgalakshanamu, op. cit., poems 7, 8 and 9 describes the great sword makers known to the author, p. 4.
29. Ibid., p. 2.
30. 'Selected Waqai of Deccan', Moughal documents translated by Yusuf Hussain, published by AP State Archaeology Department, 1955, p. 80.
31. S.K. Sinha, 'History of Medieval Deccan, History of Bijapur', Vol. II, published by AP State Archaeology Department, Hyderabad, 1964. The market prices of Aurangabad and Aurangapur in 1662, were given in the translated documents, p. 210.
32. 'Golkonda Iron' by Ishrath, pp. 252–54.
33. Ibid.
34. Irfan Habib, 'Technology and Society in Medieval India', p. 30.
35. Thevenot: 'Indian Travels of Thevenot', p. 127.
36. *Imperial Gazetteer of India.* Vol. XIII, Oxford Publication, p. 262.

17

Science and Islam

From Pluralism to Religious Essentialism[1]

S. Irfan Habib

It may sound banal to say that science in Islam was part of a pluralist project. It reached its high point during a period, which was marked by state-sponsored free thinking and borrowing from all possible sources available at the time. Nobody can dispute the fact that the early Muslim civilisation was heir to a rich and diverse intellectual stock – Roman, Greek, Indian and Persian – accomplishing a unique synthesis of ideas in all branches of knowledge. Von Grunebaum is certainly right in saying that the "stupendous rise of this civilization between 750 and AD 950 was the result of the spontaneous collaboration of the best minds of all the Empire's nationalities" (Grunebaum, 1953: 201). Abdus Salam, the Nobel laureate also had an "ecumenical picture of science" in mind when he wrote in 1988 "My own view has always been that science is the shared creation and joint heritage of all mankind and that as along as a society encourages it, science will continue to flourish in that society" (Hasan and Lai, 1984). Salam cited philosopher mathematician Al-Kindi in his Nobel Prize lecture who said "It is fitting then for us not to be ashamed to acknowledge truth and to assimilate it from whatever source it comes to us. For him who scales the truth there is nothing of higher value than truth itself, it never cheapens nor abases him" (ibid.). The scientific efflorescence of the 8th to 13th century was possible because reason and critical thinking were central to Islam. The Prophet himself left behind several of his sayings (*hadises*) where search and promotion of knowledge took precedence over religious rituals. I may take recourse to an oft-quoted saying of the Prophet where he exhorted the believers to "Seek knowledge even unto China". Let me remind the present-day Islamists that the 6th/7th century China had neither Muslims nor any religious *madarsa*, however China was the foremost centre of scientific learning. The Prophet of Islam consciously encouraged the believers to borrow ideas and expand the horizon of knowledge from all possible sources. He even said once that "he who travels in search of knowledge, to him God shows the way to paradise". He never qualified, that one has to travel only within *Darul Islam* and not elsewhere. There is no instance where the prophet insists that Quran has to be the only source of scientific knowledge. The finality being espoused today came in from 11th century onwards, when free thinkers like Mutazilites lost to the Asharites led by none other than Al-Ghazali. This led to a decisive break between the two phases of Islam – one, an early phase where eclecticism was the spirit while the latter phase was marked by closure, where inward looking Islam was projected as the true face of the religion. It is unfortunate that this latter phase is being glorified, and Islamic civilisation, including its science, is being proudly projected as a monolith, solely dependent on Quranic revelation. Today's fundamentalists take a deliberately antiquated stance: either scientific observation and theory must be made to fit the unalterable text of scriptures, or it must be shown that those scriptures anticipated modern scientific findings. Given that the Quran did not anticipate or cannot legitimate many modern discoveries, it becomes necessary to disaffirm those discoveries, and to divide science itself along cultural lines; that is, to fabricate an Islamic science consistent with the Quran in opposition to a 'Western' science unsuitable for Islamic societies because its epistemology is basically in conflict with the Islamic view (Kaiwar, 1992: 40). Suffocating from the loss of pluralism and progressive thought, which were distinctive traits of the Muslim past, how long will the global Muslim community (*ummah*) continue to suffer after the foreclosure of the gates of reasoned argument (*ijtihad*) a millennium ago? (Anees, 2001). We have reached a point where some of us are ready to disown our past; the contributions of believers are being thrown out as works of heretics. Literalism has so profoundly taken over a section of Muslim scholars that they are ready to rewrite their history of science in the light of this ill-conceived Islam.

I want to discuss the attempts of one such group called the Muslim Association for the Advancement of Science (MAAS) based in Aligarh, India. It is an irony that such an exercise is being carried out from Aligarh, a centre of Muslim reform and modernisation led by Syed Ahmad Khan in the 19th century. He aimed at turning "out free-spirited, intellectually accomplished, emotionally mature generations of young Muslims whose self-definitions would reflect liberation from the control of religious and traditional sources of loyalty and an unchallenged commitment to such underlying features of modernism as science and the ideal of progress" (Hasan, 2004: 102). Syed Ahmad countered such forces during his movement and tried hard to convey that "there is not a single nation in the world which acquired excellence, material progress, and spiritual happiness entirely by virtue of its own efforts" (Habib, 2000). Syed Ahmad felt that prejudice (*taassub*) and progress couldn't go together. Munshi Zakaullah, a close associate of Syed Ahmad and a mathematician was of the view that "the early Muslims (Arabs) did not hesitate to learn from others. They accepted non-Muslims and non-Arabs as their teachers and through their invention and researches, they expanded the frontiers of knowledge and achieved the privilege of being teachers to other communities" (Zakaullah, 1900: 2). Syed Ahmad's reformist movement from Aligarh put forth a refreshingly modern interpretation of Quranic verses, which, despite some opposition from the orthodox ulama, was accepted by Muslims by and large. The multiple readings of the Quran were much more acceptable in the 19th century than in the 21st century today. The early modernists like Maulvi Karamat Ali, who preceded Syed Ahmad, could comfortably reconcile Islam with modern science and did not see any antagonism between the two (Habib, 2004). The nineteenth century intellectuals and religious leaders confronted the situation where the evangelicals and the Company officials to undermine local religious beliefs – both Muslim as well as Hindu were using modern science. By the end of the nineteenth century they reckoned with futility of the enterprise (Habib and Raina, 1989; Raina and Habib, 2004; Gosling, 1976). In fact, the first generation of Indian scientists and intellectuals of a relatively progressive persuasion, legitimated the pursuit of modern science by neutralising the idea that modern science was Western (Raina, 2003). For doing so they argued that while science was morally worthwhile and economically beneficial, this science revealed to us the laws that governed the external world. But there was another science that the Eastern religions had grasped that revealed to us the inner world of man. This dual separation of the two realms enabled them to protect science from religion and to shield their own culture from the cultural imperialism of the colonisers (Raina, 2003).

We also need to keep in mind that most of these reformers and modernists, who espoused the cause of modern science, were not scientists themselves and quite a few were not even trained in western education. The need to acquire new knowledge, lack of which was held responsible for the colonial subjugation, became a battle cry for all nineteenth century groups except a small section of the *ulama* who called for the revival of the Islamic spiritual and ethical norms. This engagement with modern science has been dubbed as the 'colonised discourse' by some scholars (Iqbal, 2002). In a recent work, Muzaffar Iqbal goes into the early years of Islam to conclude that "unlike the Islam and science nexus that had developed naturally in the eighth century ... the new discourse is strained, laboured and carries the burden assigned to Islam in the discourse: the legitimisation of the modernists' agenda" (ibid.). There is no doubt about the fact that colonial presence conditioned most of the responses but that does not mean that it cannot be viewed on its own terms and each time we raise the issue of science and Islam we need to get back to the so-called golden age of science in Islam – almost a millennium apart from each other. It is all part of history that most of the non-European cultures were on the defensive when it came to confronting post-Enlightenment Europe and colonisation further closed its options. Even today, the post-colonial Islamic societies are faced with some real as well as perceived Western cultural and intellectual hegemonisation. This is being misused by some ideologues of Islamic science to dub modern science as part of the evil colonial baggage to be accepted at your own peril. For them modern science is an epistemological as well as cultural break from an earlier unadulterated Islamic past (Bakar, 1998).[2] For the proponents of Islamic science, modern science "is not universally verifiable for its self imposed empirical limitations put it outside the pale of a unified system of knowledge" (Anees and Davies, 1988: 249). They find it "subjective rather than objective since it is coloured with the social, cultural and historical values of the society by which it is manufactured and distributed" (ibid.). This is their critique of the Eurocentric history of science, and Islamic science is proposed as an alternative to modern science, because modern science is embedded in a particular culture and epistemology not compatible with Islamic values. This particular culture has thrown up an exploitative and unethical science and Islamic science for them is a valid alternative to this monster called modern science. It is a tragic fact of history that modern techno-science has been an active agent in the global European conquest, which has brought devastating consequences for nature as well as for colonised cultures. On this count, criticism of modern science is nothing new; rather it has grown from within science itself. However, this sectarian, religion inspired criticism of modern science rides piggyback on some valid critiques of it. The projects like the Muslim Association for the Advancement of Science is one such critique of it, which is inspired by an antipathy towards modern science, seen by them as Western science inspired by Christian ideals and values.

The 1980s saw a shift in the global Islamic politics as well as its self-perception and the major cause for this change was the 1979 Islamic revolution in Iran. For Ziauddin Sardar "the beginning of the 1980s, with OPEC power at its peak, the Iranian Revolution and a growing consciousness in Muslim societies of their cultural identity, was a period of particular optimism in the Muslim world" (Sardar, 2004). It unleashed an aggressive search for Islamic identity in all spheres of life, and also the founding of Islamic essentialist institutions and platforms in sciences as well as social sciences. Ismail R Faruqi, an eminent professor of Islamic studies at Temple University in Philadelphia, initiated a movement among Muslim scholars to Islamise knowledge. A number of Saudi businessmen came forward with a 25 million-dollar endowment to establish the International Institute of Islamic Thought (IIIT) in 1981 and the following year Faruqi published his manifesto: *Islamization of Knowledge: General Principles and Workplan*. Next step was the launching of the *American Journal of Islamic Social Sciences* in 1984. M Zaki Kirmani, who was the All-India President of the Student Islamic Movement of India (SIMI) till 1981, founded the Centre for Studies on Science (CSOS) in 1982 at Aligarh, along with those who thought `Islamically' about science and related issues. This was followed by the formation of MAAS in 1983 for cultivating similar ideas among working scientists. Kirmani's major initiative came in 1985 when he started publishing the *MAAS Journal of Islamic Science* and remained its editor till 1998. His prime motivation, in his own words was to visualise "a strong philosophical and sociological base for science, which could be developed and articulated on the basis of the Quranic thoughts". S H Nasr spoke about it in 1987 at the MIT Muslim Students Association, hailing it "as a unique institution founded by 20 or 30 scientists … . who want to absorb, first, Islamic science, then to absorb Western science well" (Nasr lecture). Nasr even declared that MAAS has been founded at the initiative of the Aligarh University, which he calls the "major Islamic university who's Islamicity is now very much threatened". Both the statements are misleading; neither MAAS has anything to do with the university nor is it an Islamic university. AMU is a secular institution, primarily aimed at the education and progress of the Muslim community, yet open to all. A perusal of some of the issues of *JIS* makes it clear that the objective was not really to absorb `Islamic science' or `Western science' but to articulate an epistemologically distinct category called `Islamic science'. An utopia, which exists only in the imagination of its protagonists. Now Sardar also seems to be disenchanted with the `Muslim optimism' of the 1980s. He had imagined that this cultural revival would work towards demanding, "The history of science in Muslim civilisation was given its due important and rightful place in the overall scheme of the history of science". However, the early nineties saw a definite shift away from Sardar's hopes and "towards more obscurantist and dangerous directions. The Islamic science discourse now follows the way of the *Taliban*" (ibid.). The issues of *JIS* are replete with articles imagining `Islamic science' as a viable alternative to modern science (Kirmani, 1990).

One of the early issues begins by proclaiming most of the contributors to the Islamic Golden Age of science as heretics, who were either *Mutazilites* (Free thinkers) or agnostics. Kaleemur Rehman further says:

> Many of them studied and practiced music ... . Al-Razi did not believe in revelations. Farabi depended on pure reason (and not *shariyah*) for discriminating between good and bad. Al-Kindi denied divine attributes. Ibn-Sina did not believe in *Maad Jismani* (resurrection of the body) ... There was gradual loss of Islamic values from the society. (Rehman, 1987: 49)

He finds it regrettable that "no two Muslim philosophers agreed with each other as far as source of knowledge was concerned. For each of them the source of knowledge was different ... ." Rehman is not prepared to concede that it is this diversity of the early Islamic civilisation, which enabled it to contribute so richly to the cumulative growth of modern science. He also refers to some of the early commentators (*mufassereen*) of Quran who committed grave errors by saying that there is a river in the sky from where the clouds collect water and bring it to the earth (ibid.). He concludes that Muslim science of al-Kindi et al., was a product of such chaotic times and information pollution, which could hardly be called science with Islamic inspiration. There is no disagreement here, the early scientists of Islamic civilisation were neither looking for anything exclusively Islamic in their science nor they had anything to do with the weird commentaries of Quran referred to above. Some of the early commentators (*mufassereen*) of Quran have done serious damage by reading science in the Quran and this exercise continues even now. However, I am more concerned about the `information pollution' caused by some Muslim scientists, not merely by commentators of the Quran, who want to harness the firepower of *jinns* to solve the energy crisis today (Hoodbhoy, 1991). The paper advises "those who are responsible for shaping the science based on Islamic values should be very careful not to base their epistemology and methodology on the examples derived from the works of medieval Muslim scientists and philosophers ... the Islamic science's structure should base from the beginning on Quran and Sunnah".

The shift referred to by Sardar had three main components and they are reflected in the papers published in the *JIS*. The first is based on the fundamentalist idea that all knowledge, including scientific knowledge can be found in the Quran. This thesis got a boost from the well-funded Saudi project on `Scientific Miracles in the Quran'. The second component can best be described as mystical fundamentalism (Sardar, 2004). And the last is the rhetoric of Islamic values. The *JIS* harps upon the Islamic ethics and morality without substantiating how that can be realised in concrete terms. Islamic science can be properly cultivated only by those inspired by the faith and values of Islam. Its exclusivism becomes obvious when Rais Ahmad approvingly cites Anees and Davies saying:

> Islamic science is an integral part of Islam as a complete way of life, the only framework within which it can be defined, it cannot be inculcated in isolation from the mainstream of the Islamic intellectual and moral landscape. Islamic science that is sub-species of Islam (and not of science) generates a world-view within the overall framework of Islamic values. (Ahmad, 1990: 26)

While talking of morality and faith, I am reminded of Jawaharlal Nehru who articulated for the possibility of morality and ethics beyond religious faith. When religion is being held up as a unique source of faith, we need to remind ourselves that there are other firm foundations upon which we can build moral and ethical projects, in both private and public life. In our own recent history, there is perhaps no better practical instance of the effort to find a non-religious bedrock for morality than that of Nehru himself (Khilnani, 2002: 1–2). Explaining his position on faith, Nehru wrote to Gandhi in 1933:

> Religion is not familiar ground for me, and as I have grown older, I have definitely drifted away from it. I have something else in its place, something older than just intellect and reason, which gives me strength and hope. Apart from this indefinable and indefinite urge, which may have just a tinge of religion in it and yet is wholly different from it, I have grown entirely to rely on the workings of the mind. Perhaps they are weak supports to rely upon, but search, as I will, I can see no better ones. (*Selected Works*, 1973: 473)

Another propounder of Islamic science S.W.A. Husaini goes a step further by rejecting the term Islamic science in favour of *Shariyyah* science and pleads that *Shariyyah* rules be applied where it is applicable in the acquisition of rational knowledge. He proposes the development of *Shariyyah* philosophy of science, *Shariyyah* history of science, *Shariyyah* sociology of science and *Shariyyah* policy of science (Husaini, 1986,). For him *Shariyyah* science meant the total exclusion of some disciplines like numerology, alchemy and music (Husaini, 1985). Contrast this approach with the nineteenth century Indian modernist reformer and maulvi based in Calcutta. Maulvi Karamat Ali, who was a teacher at a madarsa in the late 19th century and also an expert in music wrote thus in his book *Makhiz-i-Uloom*:

> Alas! That our nobility, so devoted to music, should have neglected it so far as not to carry it to that perfection when it would melt and soften the human heart. Had they taken this trouble, it would have become necessary to them to direct their attention towards mathematics. It is a matter of pain and regret that they should hear as well as learn music from ignorant buffoons. (Ali, 1867: 78)

M. Riaz Kirmani, the present editor of the journal comes out with another alternative term for Islamic science, which he calls "*ilm al-mashiyah*" (Kirmani, 1992). He finds this term closer to Islam but takes it further away from science by bringing in the role of the Creator and also acknowledging other sources of knowledge. The term *mashiyyah* means Allah's will so through *ilm al mashiyyah* Kirmani intends to carry out the study of Allah's will operative in the affairs of the universe be it nature or society. He accepts revelation as a source, which may provide paradigms for generation and application of science (Kirmani, 1987, 1990). Most of the essentialist projects are self-righteous, they deride the arrogance of modern science but drag Quran and Allah to humble not only science but also other religions. Church fought the battle with `science' on false grounds and lost the battle. The Church lost the battle because they have not the Scripture in the original form. They polluted Divine Guidance with facts against Truth (Khan, 1990: 14–15). It ends with a final statement that "we have got *al-Furqan*, the criterion (of right and wrong) which is the essence of all scriptures in original form", meaning thereby that all other knowledges and faiths are mere falsehoods. This so-called Islamic position is un-Islamic as it lacks the Islamic humility, which is the very cornerstone of Quranic philosophy as well as central to the Prophet's own conduct. The Quranic term for humility is *khushu*. The opposite of humility is arrogance (*kibr* in Quranic terminology). The Quran speaks of Satan (*iblis*) as the arrogant one who refused to obey God's command to show humility towards His creatures. In other words, one may consider the absence of humility tantamount to arrogance, which is not an angelic, but a satanic attribute. Arrogance defines its own boundaries, foreclosing new possibilities of knowledge. Further, the Quran states that arrogance leads to tyranny (*zulm*) (Anees, 2001). Besides, all this exclusivist exercise to define science in terms of Quranic revelation and Allah's will goes to undermine the very enterprise of science. In Islamic science rationality is not denied but in case of contradiction it is revelation that will prevail over rationality (Rehman, 1987: 54). In this case it is no more the pursuit of the unknown but merely the rediscovery of knowledge, including scientific knowledge, already revealed in the Quran and the sayings of the Prophet.

There is one detailed article in one of the issues of the journal, which tries to highlight the inner tensions and contrasts of the various schools of Islamic science. It begins with Maurice Bucaille and his Quran and Hadis centred Islamic science; goes ahead to Waqar Husaini and his Husainism – the Islamic science developed through imitative-innovative assimilation of modern science based on Shariyyah; next is Seyyed Hossein Nasr and his sufistic, sacred and mystical Islamic science, and the last important school of Islamic science referred to by S.B.M. Zain is the Ijmalist's view on Islamic science led by Ziauddin Sardar (Zain, 1990). All these positions can be viewed as a wish list of the Islamic enthusiasts while the author laments in the end saying "most Muslim scientists unfortunately still subscribe to or believe in this positivistic-universal-value free view on science and technology" (ibid., 70). The whole debate looks like an endless war of rhetoric among Muslims and against the West where orthodoxy and totalitarian puritanism seems to have won over reason and skepticism.

Let me end this discussion with the Nobel Laureate Abdus Salam and see how he managed to broach the Western antinomy of science and religion. Salam successfully argued from a conception that kept the scientific and religious domains distinct. Though all his scientific presentations began in the name of Allah yet it was never a constraint in his pursuit of science. Salam's biographer Jagjit Singh informs us that for Salam "... science and religion refer to different worlds; religion to the inner world of the human mind and science to the outer world of matter. To explore his inner world of 'soul' and Allah one needs faith and to explore the outer world of matter, reason" (Singh, 1992: 157). On both science as well as religion, one would agree with Salam, who was not only a successful scientist but also a practising Muslim instead of those who misinterpret both. Salam's notion of *ilm* was not merely the sacred knowledge about God but also about the profane. He rather believed in the Prophet's exhortations to seek knowledge even unto China and not like our essentialist interlocutors for Islamic science whose myopia does not take them beyond Quran and Sunnah.

A recent study dubs all nineteenth century reformist exhortations to pursue modern science as reductionist because the word `*ilm*' was conveniently used to produce a new strand of Islam and science discourse (Iqbal, 2002: 244). In that case, is it not reductionism to limit *ilm* to mean merely knowledge about God? If that was so then why did the prophet make this distinction in his famous saying "To listen to the instructions of science and learning for one hour is more meritorious than standing up in prayer for a thousand nights" (Ali, 1978: 362). The prophet of Islam exhorted his newfound followers to pursue both the sacred as well as secular and both the exercises for him were ultimately a search for the Truth of God. The concept of `*ilm*' or knowledge is being misread and misinterpreted by a large number of Islamic scholars and activists today in the same manner as the notion of *jihad* is being hijacked and trivialised by some Islamic extremists.[3] Both are akin to the vandalisation of the core edifice of Islam where *ilm* and *jihad* occupied a central place.

Notes

1. Presented at the seminar on History of Science and Technology in India, University of Hyderabad, 1–3 September 2005.
2. Even this past, perceived as unadulterated, was not really so. This most sought after and pristine Islamic past had its illustrious Nestorian Christian, Jewish, Hindu, Chinese and Buddhist contributors, who were welcomed by the liberal Caliphs of Baghdad to engage in the production of this corpus of scientific knowledge, which later came to be called Islamic science. Today one tries to forget or deliberately overlook its multi-cultural and multi-religious origins.
3. The notion of *jihad* in Islam is not so simple and easy as it is being made out to be to serve narrow and devious political objectives. *Jihad* in Arabic means to strive, which has nothing to do with violence. It is a multi-layered concept and not just one-dimensional as it is being projected. The greater *jihad* or *jihad-i-akbar* is to control one's greed and selfish desires and lesser *jihad* or *jihad-i-asghar* is war with sword in defense of Islam.

References

Ahmad, Rais. (1990). "Islamic Science: Its Scope and Purpose", *Journal of Islamic Science* 6, No. 2.

Ali, Syed Amir. (1978). *The Spirit of Islam*. Delhi (first published in 1922).

Ali, Karamat. (1867). *Makhiz ul Uloom*. Calcutta.

Anees, Munawar Ahmad and Merryl Wyn Davies. (1988). "Islamic Science: Current Thinking and Future Directions", in Ziauddin Sardar (ed.), *The Revenge of Athena: Science, Exploitation and the Third. World*. London: Mansell Publishing Limited, 249–60.

Anees, Munawar A. (2001). *From Knowledge to Nihilism: Redeeming Humility*. http://www.muslimphilosophy.com/journals/is–01/anees.htm

Bakar, Osman. (1998). *Tawhid and Science*. Lahore: Suhil Academy.

Gosling, David. (1976). *Science and Religion in India*. Madras: Christian Literature Society.

Grunebaum, Gustave von. (1953). *Medieval Islam*. Chicago: University of Chicago Press.

Habib, S. Irfan. (2000). "Reconciling Science with Islam in 19th century India", *Contributions to Indian Sociology* 34: 1.

Habib, S. Irfan. (2004). "Viability of Islamic Science: Some Insights from 19th Century India", *Economic and Political Weekly*, June 5.

Hasan, Mushirul. (2004). *From Pluralism to Separatism*. Delhi: OUP.

Hasan, Z. and C.H. Lai (eds.) (1984). *Ideals and Realities: Selected Essays of Abdus Salam*. Singapore: World Scientific.

Hoodbhoy, Pervez. (1991). *Science and Islam*. London: Zed Books.

Husaini, S.W.A. (1985). "Towards the Rebirth and Development of Shariyya Science and Technology", *Journal of Islamic Science* 1, No. 2.

Husaini, S.W.A. (1986). "Birth, Decline and Rebirth of Islamic Science and Technology", *Journal of Islamic Science* 2, No. 1.

Iqbal, Muzaffar. (2002). *Islam and Science*. Burlington: Ashgate.

Kaiwar, Vasant. (1992). "Science, Capitalism and Islam", *South Asia Bulletin*, Vol. xii, No. 2.

Khan, H.A. (1990). How to identify "Islamic Science", *Journal of Islamic Science* 6, No. 2.

Khilnani, Sunil. (2002). "Nehru's Faith", *Economic and Political Weekly*, November 30, 1–15.

Kirmani, M. Riaz. (1987). "Epistemological Foundation of Islam", *Journal of Islamic Science* 3, No. 1.

Kirmani, Zaki. (1990). "Editorial", *Journal of Islamic Science* 6, No. 2.

Selected Works of Jawaharlal Nehru. (1973). Vol. 5, New Delhi: Jawaharlal Nehru Memorial Fund.

Raina, Dhruv and S. Irfan Habib. (2004). *Domesticating Modern Science: A Social History of Science and Culture in Colonial India*. New Delhi: Tulika Publications.

Sardar, Ziauddin. (2004). *Desperately Seeking Paradise*. London: Granta.

Rehman, M. Kaleemur. (1987). "Preface to Islamic Science", *Journal of Islamic Science* 3, No. 1, 45–56.

Singh, Jagjit. (1992). *Abdus Salam*. New Delhi: Penguin Books.

Zain, S.B.M. (1990). "Modes of Operations in the Quest of Islamic Science", *Journal of Islamic Science* 6, No. 2, July-Dec.

Zakaullah, Munshi. (1900). *History of the Physical Sciences of the Muslims* (in Urdu). Delhi.

18 The Needham–Hora Controversy

Disagreements over the Chronology and Approach to the History of Science

Dhruv Raina

The evolution of disciplines has been a subject of recent interest in the social sciences. In particular, a subject of interest has been the migration of disciplines across cultural boundaries and their subsequent redefinition in assorted educational, political and pedagogic environments. In this paper, I examine a controversy around the domestication and globalisation of the history of science in the decade following the attainment of independence from the colonial rule in India. In particular, this controversy is woven around the attempt to institutionalise history of science as a discipline and the controversy rages around the themes of historiography, the nation, institutional boundaries, and intellectual collaboration. In other words, it is about how a set of norms, methods and practices accepted by a global community of historians of science was adapted in the Indian environment and the negotiations that are consequent to altering some of these norms. The controversy is important, because it opens a window in India, throws light on the historiographic frames and actors that set the agenda for history of science in India between 1950 and 1980; and finally, because it reveals the networks of international collaboration that enabled Indian scholars to join the global project of producing a new history of science in non-Western societies. The essay is itself part of a study of the triangular relationship between members of Gary Werkey's 'Visible College' (Werskey, 1978), UNESCO, and the National Institute of Sciences of India (later called the Indian National Science Academy)

In an earlier paper I had set out a map of this collaboration between the 'Visible College', UNESCO and members of the Indian scientific community and identified some of the actors (Raina, 2005). As suggested above, this paper looks at a controversy surrounding the Proceedings of the Symposium on the History of Science in South Asia organised by the National Institute of Sciences, New Delhi and the South Asia of UNESCO. The symposium was held in the rooms of the University of Delhi from November 5–7, 1950. Scientific controversies, as we well know, are appropriate units for studying science (Engelhardt and Caplan, 1987), and I shall venture into the disciplines in the making. And the one that shall be discussed is not only one about historiographic differences but about the theory of knowledge and divergent conceptions of what an interdisciplinary history of science meant. The controversy never resonated within the community. In other words the community of historians of science in India found a way to proceed with their own research without seeking a resolution within the international community, which revealed the weak ties with the global community of historians of science. This also says something about the nature of history of science as a discipline, but more about that later.

The National Institute of Sciences of India was a scientific society that was later to be renamed the Indian National Science Academy; and one of the objectives of the former was "to promote and maintain a liaison between science and letter"[1] (Symposium, 1952: 323.) And the symposium on the History of Sciences in South Asia organised by the National Institute reflected an interest of the institute in maintaining such a liaison but signalled in the 1950s, the decade of the most rapid institutional development of sciences in post-colonial India (Raina and Jain, 1997), the intent of reinventing and configuring an identity suitable to the needs of the modernising Indian State. Metatheoretically speaking we can nevertheless observe that the need for maintaining such a liaison arose out the perceived disconnection between the two realms of the sciences and letters. In fact, in the understanding of the network of scientists associated with the National Institute of Sciences of India, the development of the sciences may not augur well for society in the long run and hence the need for a humanist garnishing (Symposium, 1952: 323). Thus, the Symposium was planned as a "first attempt (in India) to bring together historians, oriental scholars and scientists" to address some of these issues, and on the specific subject matter of history to resolve central concerns of what constituted source material for the history of science in India, and more importantly to agree upon a chronology that would be found appropriate to the two camps.

It was the noted physicist from Delhi University, D.S. Kothari who first mooted the idea of holding a symposium on the History of

Science in South Asia; and thought that it was appropriate to involve UNESCO and he subsequently raised the matter with Professor Alexander Wolsky, a Hungarian biologist who was Principal Scientific Officer at the UNESCO Science Co-operation Office for South Asia, in New Delhi, between 1948 and 1954 (Symposium, 1952: 324). He later went on to teach experimental embryology and genetics biology at Fordham University and retired from Marymount College in the United States (Tarateta, 2005). This further explains his being associated with the Needham, the former embryologist. Kothari's proposal was welcomed by Wolsky, ensuring the former of the National Institute to initiate and organise the Symposium; UNESCO's Science Co-operation Office would support it financially (Symposium, 1952: 324). A programme was then developed and discussed by the two of them. The subsequent report of this meeting suggests that his was to be no ordinary Symposium, even though a sponsoring Committee and a convenor were to be appointed to overlook the actual organisation of this Symposium. To ensure that the Symposium was an intensive meeting that would survey the existing state of knowledge in the history of science of South Asia, it was decided that the Symposium be structured as a study group with no more than 25 participants. And one of the objectives of the survey would be to "correlate the advance made in different branches and the social implications of the scientific developments" (Symposium, 1952: 324). Underlying this objective was not just the hope of chronicling the manner in which the advance of the sciences in India had contributed or otherwise to social development.

The hope was that the Secretariat of the International Union for the History of Science would be requested to provide support in two forms. The first was logistics in the form of background reading material and working papers that would guide the scholars into the contemporary research concerns and methods of the history of science. Further, it was suggested that a leading historian of Science be invited to attend the Symposium as an 'expert consultant' (Symposium, 1952: 324). Wolsky informed Kothari that he had already written to Needham informing him that a meeting was being organised on the History of Science in South Asia and requested Needham to attend the meeting scheduled for November 1950. He was also apprised of the scholars likely to attend the meeting.[2] Both Needham and Bernal were scientist-historians of science, who had emphasised the importance of the history of science for the current practice of science (Fuller, 1999). Further, they were both acquainted with India and the freedom struggle and were members of a 'Visible College' that was well acquainted with India's freedom struggle.

Between 1947 and 1950, Bernal had already made at least three trips to India. The long and short it, was the Symposium was intended to seed and institutionalise the history of science in India on lines very similar to that of professional history of science in the West. This not only meant acquainting the Indian community of historians with the new sets of methods and techniques, but more importantly with identifying and emphasising the current research priorities in the history of science at the international level. It appears that the prevalent idea was that the mode of establishing and institutionalising scientific disciplines was applicable to an interdisciplinary field such as the history of science.

The intent then was to develop the discipline of the history of science. In any case, the point must not be lost. If the history of science was about the modes of disciplinary institutionalisation, then here was history of science at work; where it became the theory and mode of its own institutionalisation. At least that is the way Needham and Wolsky in UNESCO thought about the matter. Thus, every delegate at the Symposium was required to submit a report on the history of sciences of particular branches in their country. This report was to contain a listing of published material that could inform the researches of scholars undertaking 'intensive study' (Symposium, 1952: 324.) It is interesting to observe the number of times the term 'intensive study' or "intensifying the study of the History of Science in South Asia" appears in the discussions preceding the meeting. This clearly reflects on the current amateurish status of the history of science was underdeveloped in the region at the time. A few scholarly works had been published by scientists such as P.C. Ray, A.N. Singh and others; but this was not enough to build a discipline or did not intersect with concerns and questions that were being investigated by leading historians of science in the field at the time.

These limitations, if that is the way we choose to label them, were recognised by Kothari and Wolsky in their discussions. In planning the conference they grouped the participants into historians, scientists, and research scholars working on the history of science who would separately formulate practical recommendations (Symposium, 1952: 324). The recognition of the boundaries constituting South Asia are interesting. Thus, it was thought necessary to invite delegates from Thailand, Ceylon, Burma and Indonesia. Pakistan, which then included Bangladesh, was left out. The bitterness of partition in an academic seminar. Nevertheless, as the idea developed it was decided to send a circular to all the countries mentioned, including Pakistan seeking their cooperation and contributions from historians and scientists (Symposium, 1952: 326). The historian of science is clearly absent from the list of contributors and the conference was seeking to create that new profession. In any case, it was hoped that the host of measures proposed would enable the participants to put up a good show at the Congress of the International Union of the History of Sciences (IUHS) that was held in Amsterdam in August 1950 (Symposium, 1952: 325). The terms of reference of IUHS, its charter of activities that were directed at broadening the outreach of the history of science globally of the Government of India and its Department of Scientific Research to collaborate with UNESCO in organising the symposium (Symposium, 1952: 326).

The sponsoring committee met twice that year – in March and September 1950, where the theoretical assumptions and a tentative programme were laid out. The period of interest was charted as 200 BC to 1857 BC; namely from the Indus Valley Civilization to the beginning of British rule. The focus of the history of science in this first schematisation was on chronology, achievements, biography of scientists; transmissions were concerned that they were treading on very delicate ground, for the euphemistic language in which the theme was articulated betrays a nationalist sensitivity that has not evolved into a cross cultural or comparative history. The social relations or context of science while present was marginal to theorisation of the history of science. Social history appeared as 'general history' with no ties to the history of science, but the impact of technology on social development was outlined in a little more detail. The last two sections were to be devoted to the development of logistics and organisational matters. This should indicate to us the state of history of science in India at the time. There were certainly no chairs or positions in universities or research institutes at the time. Bibliographies or catalogues or catalogues on the History of science were not available or had not been compiled and certainly source materials and libraries had still to be identified (Symposium, 1952: 325–6). The structure of the seminar clearly reveals that the purpose of establishing both a cognitive and institutional identity and an institution was foremost in the minds of the Sponsoring Committee.

The Symposium was finally held between the 5–7 November 1950 in the rooms of the University of Delhi, with hospitality being offered by the Vice Chancellor and the Dean of the Faculty of Science of the University. There were 30 participants that included representatives of 12 learned societies (Symposium, 1952: 326–7). The spirit of the Symposium appeared to reaffirm UNESCO's commitment to the history of science as embedded within the history of civilisations. The official report of the symposium celebrated two features of the Symposium: (1) that for the first in India "trace the History of Sciences", (2) that there was a general consensus and human mental evolution. Consequently, this history could only be traced though the collaboration between these two scholarly communities (Symposium, 1952: 327). While this consensus was officially proclaimed it did not manifest itself in the papers presented. All was not hunky dory. The organisers had anticipated that there would be disagreements about chronology and fixing the dates of important ancient scientific texts, especially those dated to the pre-Christian era. Consequently, a Chronology Committee[3] was appointed that agreed upon a table that would serve as a working chronology. This table drew upon the chronology of Indian texts presented in Winternitz's History of Indian Literature (Symposium, 1952: 327; cf also vide Appendix III: 331). Of the thirty participants there were four historians who had already done research and published extensively in their respective fields of expertise. The rest were scientists and mathematicians. A.N. Singh, a mathematician had authored a book on the history of Indian mathematics with B.B. Dutta. That was a standard reference book on the subject for several decades. There were a number of young scientists present there who would later on go on to publish and write a number of books and research papers on the history of science. These included A. Rahman, who was then working for the Council of Scientific and Industrial Research, K.V. Shukla, a mathematician who went on to author several books on the history of Indian astronomy and translate the canonical astronomical texts of the tradition from Sanskrit to English.

The organisation of the session was noteworthy for it indicates the subjects considered important and the historiographic frame within which the organisers were working. The symposium was divided into five sections, the last of which was on general problems, methods and philosophy of science. The sequence of events led naturally in a positivist chain from astronomy, physics, geography and geology to mineralogy, pharmacy and biology to applied science, technology and engineering. The organisation of the session and order of their appearance was an index of the dominant theory of science, with matters of technology, social history and history teaching featuring very low in the order of things.

Twenty-four papers were presented in all, of which four dealt with the chronology of scientific disciplines in India, and four were about India's "achievements in" or "contributions to" again, specific scientific disciplines. This clearly indicates the perceived urgency in the post-colonial nation for constructing a scientific heritage and genealogy to the present. Rahman presented two papers in the general section. The one on "Social Factors in the History of Sciences in India", and another on "The History of Sciences and Some Problems of teaching" (Symposium, 1952, vide Appendix VI: 334–5). The tone of both these papers was different from the rest, both in terms of the historiography proposed and in terms of setting up a curriculum for teaching the history of science.

Dr S. Hora opened the symposium with the remarks that history was a useful guide in forecasting the "future of a nation". The task of the history of science was to establish the foundations of the "traditional knowledge and achievements of the nation", on which to rest the future development of science (Symposium, 1952, Appendix V: 333). This could be done by sifting and evaluating scientific achievements of any country "... according to modern standards ". This presentist reading would then serve as an "index to its past glory". Methodologically, then for Hora, contemporary science provided the framework to interpret the past. The results of which would serve not merely the cultural assimilation of modern science, but more importantly inform the technological development of the country. Interestingly, he aligns the trajectory of the symposium with UNESCO' objective to achieve "unity among warring nations in the study of science" This could be accomplished if the study of the history of science were to be "free from personal or racial prejudices ... any nation can be rightly proud of its past achievements, but it should not be oblivious of the achievements of other nations" (Symposium, 1952, Appendix V: 333). The symposium was thus inaugurated in the spirit of scientific internationalism regarding which the sponsor, namely UNESCO, and the local organising committee appeared to be deeply in agreement.

Needham's paper on the "Unity of the Sciences" was circulated among the participants (Needham Correspondence: Wolsky to Needham, 15/1/1951) and its presence was reflected in the opening address of Hora. A set of the papers presented was later sent to Needham. The proceedings of the Symposium were published in *Nature* in April 1951. Needham however, wrote a very scrupulous anonymous review of the papers and send it to *Nature*. On 17 May 1951, *Nature* wrote back to Needham concerning his review and remarked that they "cannot very well be asked to take the responsibility of printing such an article without having a signature appended to it" and they sought his permission to include his name as author of the review. In the meanwhile Wolsky had published his own review of the proceedings in the Archives Internationales d'histoire des sciences (1951).

Needham as I said earlier had not attended the conference but the UNESCO office at New Delhi had sent him the papers in a mimeographed form, which comprised about 180 pages of single space, mimeographed foolscap-size typescript[4] (Needham, *Nature,* vol. 168, July 14, 1851, footnote page 64). He began by highlighting the importance of the history of science of India and other part of South Asia for the "general cultural history of mankind". This emphasis was significant from the UNESCO perspective of the history of science and culture.

Further while recognising that a great deal of work had been done that was available for synthesis; which was an euphemistic way of saying that the material required to be refracted through a robust historiographic frame, wherein the development of science and scientific thought in India needed to be embedded within a comparative perspective of parallel developments in Europe and other parts of Asia (Needham, 1951; *Nature,* 64). However, at the very outset Needham picked on the issue that had been identified by the Chronology Committee and going by the discussion at the Symposium and Needham's review, the Symposium proceeded, quite oblivious of the recommendations of the Chronology Committee.

The problem appears to have arisen from the different methods applied by professional historians and professional scientists turned amateur historians of science in dating important historical manuscripts, theories and personalities. The problem of exaggerated chronologies that Needham perceived were not to be found in the works of historians such as A.S. Altekar and R.C. Majumdar, but were those relating to texts dated to the first two millennia of the pre-Christian era. He particularly identified the papers of Shukla and Dixit on Indian Astronomy, that of the chemist N.R. Dhar on Indian Chemistry, R.V. Seshaiya on Embryology which happened to be Needham's own specialty, and G.V. Satyanarayanamurti on medicine (Needham, *Nature,* 1951: 168, 64) Needham went to draw up a comparative table of dates of Indian datings of the Upanishads, the Jotisa Vedanga, the Surya Prajnapti, Surya Siddhanta, Arthasastra, Susrutha Samhita and Caraka Samhita. And found that K.R. Dixit, K.S. Shukla, P. Ray, S.L. Hora and R.V. Seshaiya dated these texts from the pre-Christian era to anywhere between 2000–4000 years earlier than they were by professional historians like Majumdar and Altekar. In fact, there was a divergence of opinion ever amongst the scientist-historians (ibid.). Needham did of course, not get an opportunity to see the text of the discussion of the various sessions, wherein both Altekar and Majumdar had cautioned these amateur historians of science from such exaggeration.

In Needham's reading three historiographic features were shared by the papers he had identified. The sexagesimal division of the circle and the system of twenty-eight lunar mansions travelled from India to the Babylonians. They all betrayed a "chauvinistic tendency" that minimised foreign influences on Indian science, and placed greater emphasis on outward transmission. The task was simplified since the chronology was still debatable. The third was the absence of the comparative method. These were further compounded by a deep scientific presentism which involved "reading too much into the ancient texts" in the image of contemporary science (Needham, *Nature,* 1951: 168, 64–5). Thus, faulty chronology facilitated a chauvinistic history, whose tenuous claims could not be checked in the absence of a comparative perspective of history. These put together with scientific presentatism produced a contestable history of the sciences of the South Asian region. The central problem then was to settle the chronology of the Indian texts, for this would enable the resolution of issues relating to the transmission of texts, theories and ideas (Needham, *Nature,* 1951: 168, 66).

With this clarification Needham proceeded to reiterate the importance of the history of science in the South Asian region. The way out of this double bind of Eurocentrism and an incipient Indocentrism, Needham finds it in the project of a kindred soul, the French Indologist, Jean Filliozat. Quoting Filliozat extensively, Needham gets Filliozat to put a finger on the dilemma.

> Indian scholars, moved by national pride and prone to maintain that their sciences in high antiquity surpassed even that of today. In the west, on the other hand, many maintain that the spirit of scientific research could only have been born in Europe, and that what science the Indians had, they borrowed. (Filliozat, quoted in Needham, *Nature,* 1951: 168, 65–6)

This was the standard trope of the history of science as it became institutionalised in the West in the nineteenth century (Harding, 1998; Raina, 2003a). There was a typical asymmetry in the argumentative styles of history of science scholarship that both Needham and Filliozat were trying to undo in their attempt to develop a more cognitively just account. As Filliozat wrote:

> Those who speak with the greatest certainty in these matters are just those who are familiar with only one of the two traditions, knowing the other one only by scattered facts, or studies which they are unable to appreciate. To say nothing, of course, of those authorities who know nothing of either of the two traditions about which they speak. (Filliozat, quoted in Needham, *Nature,* 1951: 168, 65–6)

The way out of the perplexity suggested by Filliozat was to embark on a profound study of "Indian scientific developments in parallel with those that took place elsewhere about the same time". Such a study would reveal two things; the originality of the sciences involved, and India's and the region's true contribution to the growth of scientific knowledge (Filliozat, quoted in Needham, *Nature,* 1951: 168, 65–6).

Five months later Hora as president of the National Institute of Sciences responded to Needham's review of the Symposium papers (Hora, *Nature,* vol. 168, December 15, 1961: 1047–8). Hora's strongly worded by historically weak response, began by pointing out that Needham had not quite gauged the spirit of the conference at all; and proceeded to outline the evolution of the idea and the nature of the papers presented. Needham was well acquainted with these details and his own review had raised these very points in the introductory paragraphs. Hora went on to identify a fundamental issue of historical issue of historical method as far as the history of sciences of the South Asian region was concerned. The scientists-historian of science working on ancient materials and documents had to base himself on the interpretations of orientalist scholarship. But these oriental scholars in Hora's view lacked the ability to appreciate and evaluate the "scientific thought of ancient India".[5] Hora felt that it was only for scientists "to make a correct approach to and appraisal of the scientific knowledge of ancient India", and it was their moral duty to do so (Hora, *Nature,* vol. 168, December 15, 1961: 1048). He thereby missed out addressing any of the other historiographic remarks that Needham had put up for discussion. Further, Hora adopts the authoritarian voice of science, rather than its discursive voice, and in the process signals the irrelevance of historians or oriental scholarship to the history of science. This was in total contrast to the interdisciplinary programmes that were emerging in leading schools of the history of science throughout the world, though the Cold War climate had given some legitimacy to the internal-external distinction that had begun to harden and rule a little longer.

Needham did not push the discussion any further, but remarked that Wolsky's review that appeared in the Archives Internationals was in concordance with Needham's own observations (Hora, Nature, vol. 168, December 15, 1961: 1048). It must be said in all fairness that the National Institute of Sciences of India reproduced the papers: Needham's review in *Nature,* Hora's response to Needham's review in the Proceedings of the Institute. Needham did not possibly see the discussion that followed the presentation of each of the papers. Were he to have followed that discussion he would have noted happily that several historians had raised just the points he had raised in the review; and that the issues he had taken were larger historiographic and theoretical concerns that the scientific community was quite oblivious of, concerned as it was in creating a new modern identity for itself by tracing a genealogy to the hoary antiquity of Indian science, whose glory surpassed the present. This nationalism was radically different from that of the founder of the discipline of the history of science in the last decades of nineteenth and the early decades of twentieth century.

The fallout of this controversial exchange is difficult to gauge. But it came to mean different things to different actors. As far as Needham was concerned he remained in touch with those historians of science in India with whom he shared a historiographic frame and concerns. He remained quite distant to the larger community. This can only be inferred from his sedulously maintained correspondence. Wolsky wrote to Needham in January 1951 having learned that Needham's wife was recovering in Southern France. Wolsky felt that Needham's own paper was appreciated and that the environment for the history of science had livened up in Delhi University after the Symposium. Rosenfeld had been invited from Manchester and Rahman from Hyderabad to give a series of lectures on the history of sciences (Needham Correspondence: Wolsky to Needham, 15 January 1951). Wolsky was still hopeful that Needham would come to India and bless the development of the field. Shortly afterwards Wolsky left New Delhi for Djakarta. Rahman who maintained contact with Needham till Needham's death wrote him in January 1951: "Your review in Nature of the Symposium on History of Science held at Delhi has aroused a controversy, which unfortunately has taken a chauvinist turn" (Rahman to Needham, 25 January 1952). Rahman rightly recognised that the main point of the Needham's review was "to emphasise the synthetical approach to the problem in order to crystallise a pattern of development: (Rahman to Needham, 25 January 1952). Rahman met Hora at the Calcutta Science Congress and broached the matter with him and learnt that Needham accorded preferential treatment to Chinese inventions over Indian ones. In the assessment of the Symposium, Rahman suggested that the symposium had not resulted in "the beginning of research on History of science in India, not to say of the systematisation and crystallisation of research". This is not to say that it was not likely to happen in the future (Rahman to Needham, 25 January 1952).

In fact, the National Institute would take some radical decisions to get some projects going by 1953–54, that I have discussed elsewhere and will not bore you by repeating myself here (Raina, 2003). In developing these projects, lip service was paid to the Needham's main approach and the Needham's main project; but whether the approach was actually deployed or the project redefined is again a separate matter. The synthetic project developed by Lucien Febvre, Marc Bloch, Joseph Needham and Julian Huxley within the confines of UNESCO (Petitjean, 1999) certainly did not take off in India, the reasons for which lie in the historiophic divide that this essay has attempted to trace. Every now and then the absence of the Needham's main version of the SCI would be discussed and written about. This would justify further financial support for the history of science for similar projects, signing the glory of Indian science and civilisation. What would emerge would be a construction so dissimilar and unsatisfactory that cultural studies of science could certainly make it a subject for investigation.

Notes

1. An extended report on the Symposium with a shortened version of the papers and discussions, as well as several appendices as "Symposium on History of Sciences in South Asia", Proceedings of the National Institute of Sciences of India, Vol. XVIII, 4, 1952, pp. 323–62. I shall hereafter refer to this document as Symposium, 1952.
2. Needham papers at Needham Research (hereafter NRI), Cambridge, (Correspondence: Wolsky to Needham, 16/2/1950).
3. The committee was chaired by the historian R.C. Majumdar, and the other members included A.S. Altekar, P.C. Bagchi, S.L. Kothari and A.N. Singh (Symposium, 1952, p. 331). Bagchi and Majumdar were the only trained historians among them.
4. Needham Correspondence: Letter from *Nature*, 17/5/1951.
5. P.C. Ray had made this observation about Barthe's history of Indian chemistry in the early years of the twentieth century, of the chapter on P.C. Ray and Berthelot in (Raina, 2003).

References

Engelhardt Jr., H. Tristram and Arthur L. Caplan (eds.) (1987). *Scientific Controversies: Case Studies in the Resolution and Closure of Disputers in Science and Technology*. Cambridge: Cambridge University Press.

Fuller, Steve. (1999). "Towards a Prolegomena for a Global History of Science", in S. Irfan Baib and Dhruv Raina (eds.), *Situating the History of*

Science: Dialogues with Joseph Needham. New Delhi: Oxford University Press, pp. 114–151.

Petitjean, Patrick. (1999). "Needham, Anglo-French Civilities and Ecumenical Science", in S. Irfan Habib and Dhruv Raina (eds.), *Situating the History of Science: Dialogues with Joseph Needham*. Delhi: Oxford University Press, pp. 152–97.

Raina, Dhruv and Ashok Jain. (1997). "Big Science and the University in India", in John Krige and Dominique Pestre (eds.), *Science in the Twentieth Century*. Netherlands: Harwood Academic Publishers, pp. 859–77.

Raina, Dhruv. (2003). *Images and Contexts: The Historiography of Science and Modernity in India*. New Delhi: Oxford University Press.

Raina, Dhruv. (2003a). "Betwixt Jesuit and Enlightenment Historiography: The Context of Jean-Sylvain Bailly's History of Indian Astronomy", *Revued' Histoire de mathematiques* 9: 101–53.

Raina, Dhruv. (2005). "Finding a Home for the History of Science in Post-Colonial India: The Influence of Needham and the Role of UNESCO", in Arun Bandopadhyaya (ed.), *Science and Society*. Manohar Publications, 2005.

S. Harding. (1998). *Is Science Multicultural? Post-colonialisms, Feminisms and Epistemologies*. Indiana University Press.

Tarateta, Maria. (2005). "Longtime Professor at Fordham and Marymount Dies", http:/www.fordham.edu/Campus_Resources/Public_Affairs/Inside_Fordham/Inside_Fordham_Archi/February_2005/In-Brief/

Werskey, Gray. (1978). *The Visible College*. London: Allen Lane.

19 Building Interdisciplinary Scientific Specialties

A Case of the Solid State and Structural Chemistry Unit, IISc, Bangalore

Debasmita Patra, E. Haribabu and P.K. Basu

For a sociological study of the formation of a scientific specialty and to understand its dynamics, historical sensitivity is very important. Similarly, for a historical study of science, sociological insights are needed in relating science to its context. The present paper is based on a study of an interdisciplinary specialty and institutionalisation of its cognitive and organisational dimensions in India. The paper attempts to identify the factors that contribute towards the formation and development of a scientific specialty, that is, the Solid State and Structural Chemistry Unit (SSCU) and its practice in different contexts. The present study examines how for the first time in India, the SSCU as a department, exclusively devoted to teaching and research of solid state chemistry was established in 1976 in one of the premier science institutions of India, that is, IISc (Indian Institute of Science), Bangalore. This study takes into consideration twenty-five years since the establishment of the SSCU, that is, from 1976 to 2001, to understand the dynamics of the research agenda of the scientists and their national and international linkages in terms of collaboration. Analysis of the published works of the scientists of the SSCU over these twenty-five years yielded some quantitative measures to discern the direction of research and the pattern of collaboration. Further, the scientists in the SSCU as well as some of their peers located in other organisations were interviewed to understand the micro-level processes relating to interaction and decision-making with regard to setting the research agenda and leadership both at the cognitive and organisational levels. The major findings of the study indicate that the scientists, who were recruited with specialisations in different sub-areas of solid state chemistry and of physics, brought in different theoretical and methodological skills that led to setting the research agenda of the Unit and in establishing collaborative partnerships with scientists in India and abroad. The charismatic style of leadership of the founder of the Unit and his managerial abilities helped in mobilising cultural and physical resources to facilitate the research activities of the scientists. The pattern of collaboration suggests that the SSCU scientists, during the period under study, had more collaborative partnerships with scientists located in the advanced countries than they had with scientists in India, except within the IISc.

Social/Cognitive Bases of Development of Scientific Specialties

To understand the dynamics of the development of a specialty, it is important to trace the social conditions, both internal to science and external to science, in which a specialty emerges. The sociology of science is a specialty that focuses on how and in what ways social factors both internal to the world of science and external to the world of science influence science. Earlier, the distinction between internal and external factors was rigid. But gradually the gap separating the internal and external factors of science started withering away. As Bloor (1976, 1984) said, "all knowledge, including scientific knowledge is socially and culturally conditioned". Restivo (1998), an electrical engineer by training, points out that the term social is not only in the external social and cultural milieu or context of science, but is also present in the internal social organisation of science, indeed in scientists themselves. Social studies of science have emphasised that the cognitive aspects of science must be included in its empirical investigation. These studies also pondered upon how the social and cognitive components interact in the production of knowledge, and in the relationship between different forms of cognitive (scientific) knowledge and society (Cetina, 1981). Latour and Woolgar (1979) have pointed out that the society/cognitive distinction prevents the social scientist from examining its role within scientific activities themselves. Distinctions between the cognitive and the social, the technical and the career-relevant, the scientific and the non-scientific are constantly blurred and redrawn in the laboratory. Furthermore, traffic between social and technical or scientific areas is itself a subject of scientific negotiation: today's socially produced knowledge claim may be tomorrow's technical scientific finding, and vice versa (Cetina, 1981).

The sociology of science is a relatively new specialty that focusses on the field of science studies. The development of sociology of science as a specialty within sociology, itself shows the factors, which play major roles, in the development of a specialty. However, in the present paper we are dealing with the institutionalisation and development of a scientific specialty, that is, solid state chemistry in India. Also a brief discussion about the concept of 'research area', 'specialty', etc., helps us in understanding the conditions which contribute toward the formation of a scientific specialty.

Conceptualisation of Scientific Specialties

Specialisation is the hallmark of modern science. The study of scientific specialties by sociologists of science is particularly self-exemplifying. During the late 1960s and all through the 1970s, much research focused on conceptualising what a specialty is, its components, its dynamics. Hagstrom (1970) characterised specialties as "microenvironments for research, as frequently traversed regions on a blurred map of science". He writes further: "the set of scientists in a discipline who engage in research along similar lines can be called the scientific specialty". It is reasonable to believe that scientists will communicate most often and interact intensively with fellow scientists in their specialties, exchanging preprints with them, citing their work and exchanging reprints. There is some evidence that specialties are not invisible colleges or tightly knit networks of communication, yet most specialties in science are quite small, small enough so that they could be tightly knit communication networks, even if they are not so in fact. In one of the earliest accounts of the emergence of new specialties, Crawford (1971) describes how one specialty, sleep research, is effectively organised as a communication system around certain key individuals and research centres. Griffith and Mullins (1972) have argued that major changes in a variety of disciplines have been generated within small, socially coherent groups. Crane's (1972) work on specialties in rural sociology and mathematics has led naturally to the idea of 'mapping' the structure of specialties. Holton (1972) notes that, specialties arise out of scientists' intense competitiveness and interest in making new discoveries. Mullins (1973) defines a specialty as "an institutionalised cluster which has developed regular processes for training and recruitment into roles which are institutionally defined as belonging to that specialty". A scientific specialty consists, according to Small and Griffith (1974), of a group of scientists working on similar problems and regularly communicating with one another. Whitley (1975) states that the cognitive components of specialty consist of the research practices, techniques, explanatory models, specialty concerns and metaphysical values or beliefs which underlie an area's research activities. Specialties have been identified as the building blocks of science. Specialty formation is, therefore, taken as a 'strategic research site' for examining the connections between the growth of scientific knowledge and its social organisation since new lines of inquiry are developed and elaborated in the social context of specialties (Zuckerman, 1988). Kadushin (1968) shows that workers in scientific specialties are organised in a 'social circle'. Such groups have been identified with Price's (1963) notion of 'invisible college', and have also been seen as the social location of 'technical and cognitive norms' (Mulkay, 1980).

A specialty also draws upon cultural resources – ideas, theories, models, techniques – from other disciplines/specialties. This creates conditions for the emergence of a division of labour among scientists in a specialty, who bring diverse kinds of expertise. Law (1973) by following Durkheim's notion of division of labour defines mechanical and organic solidarity in science. In science, mechanical solidarity may be defined as the development and maintenance of relationships which depend on shared standards and exemplars, and hence on a relatively high degree of consensus about theory and method. Whereas, organic solidarity may be defined as an aspect of the division of labour in which scientists come into relationship with one another because one performs services, which the other can't easily carry out for himself/herself. He takes up a case of the development of a scientific specialty namely, X-ray protein crystallography to substantiate his views. In the thirties, there was a crystallographic community, which included scientists who chose to work on proteins. For a period of time, protein crystallographers were in contact with other protein workers who were not crystallographers. Initially, many meetings were arranged with scientists from many different disciplinary backgrounds like physics, biology, cytology, chemistry, genetics, embryologist, mathematics, philosophy, zoology, etc. The group which was working on proteins in the 1930s expanded by the addition of students. In the 1940s a number of crystallographers worked on proteins, but in different ways and with differing emphases. They formed what is called a protein community.

Stages in the Growth of Specialties

In the early stages of growth of a research area, an interesting cognitive event for example acquisition of a paradigm (Kuhn, 1964) occurs and attracts new scientists to the area. This is followed by a period of cumulative development of knowledge (in terms of number of publications and new authors). Price (1963) has shown the stages of growth through a logistic curve. According to him, after a relatively slow start, disciplines exhibit a period of exponential growth followed by a period of linear growth and finally by a period of slow and irregular growth. At the beginning stage, a new area is ill-defined, both socially and intellectually. At this stage social contacts are minimal and those at work in the emergent area only gradually become aware, as preliminary publications begin to appear, that others share their interests. Stage one, therefore, is characterised by open competition, vigorous controversy and pronounced fluctuations in the support for the various perspectives present in the area.

In the second stage, which may be akin to 'normal science period' of Kuhn, regular informal relationships emerge among the scientists through which an effective division of labour is established within the network, thereby reducing that overlap of research interests which fosters competition. Research problems become increasingly well defined; consensus on major scientific issues – both theoretical and methodological – becomes increasingly apparent. In most areas the elevation of the scientific leaders is parallelled by that of a few men whose contribution seems to have been at least as much organisational as scientific. Such organisational leaders tend to excel in promoting the field within the research community and sometimes among the general public, in attracting funds and in building up research groups and collaborative enterprises. Thus, the research network tends to become socially institutionalised and scientifically routinised. The third and final stage is less predictable than the previous two stages. As the scientific value of the range of allowable topics declines, so the network is increasingly likely to lose existing members and to perform poorly in the competition for young researchers. The growth, decline and branching of several streams of related research areas produce the exponential pattern of growth which is the characteristic of major specialties or disciplinary areas (Mulkay, 1980).

Types of Specialties in Science

Law (1973) has distinguished three different types of specialties in science. These are as follows:

1. A technique or method-based specialty that constitutes an interacting group of scientists, whose solidarity rests on the basis of shared scientific gadgetry, and its development.
2. Theory-based specialties are defined in terms of a shared formalism – members are those whose main concern is theory and

its development, and exemplars relating to various gadgetries and problems that arise out of the central concern with theory.

3. Subject-matter specialties have members who work on a particular subject matter or problem. Members of such a specialty are prepared to use a variety of techniques and theories, none of which may be the preferred one, in general.

Solid state chemistry, the object of the present study, falls under the third category. Because the members of the solid state chemistry community work on various problems using a variety of theories and methods, none of which may be the preferred one, in general as mentioned above.

From the above sociological literature concerning the nature of scientific specialties and their concerns, we shall examine certain conditions which contribute towards the formation, growth and sustainability of scientific specialty.

Conditions Leading to a Scientific Specialty Formation

Shared Interdisciplinary Research Problems

The central component around which a specialty emerges is a research problem that cannot be solved with conceptual experimental apparatus of a given discipline. If the problem can be solved within an established disciplinary set-up, not a single member will look out for another discipline for ideas. So, these interdisciplinary problems draw attention of scientists from different disciplines who are equally interested with the same kind of problems, thus, creating conditions for the emergence of a specialty.

Organisation

An organisation is an association of human beings deliberately constituted to accomplish a goal. Organisation performs many things for the development of a specialty. It organises manpower to bring scientists together who are working in the same area; organises physical infrastructure; facilitate use of funds on the basis of appropriate norms; an intellectual ambience for interaction among scientists within their groups and across groups (Beardsley, 1964).

Funding

Funding is essential for any specialty to carry out research agenda. Generally there are two major sources of funding: the governmental and the non-governmental. In sciences, the non-governmental sources basically refer to the industries. In some cases it is found that in the infant stage of the growth of a specialty, some individuals contributed a major part of the funding towards the development of the specialty.

Publication of a New Journal

Journal is a medium through which the scientific communities share their methods, theoretical models, findings, etc. In the formation of a specialty a journal plays a major role to make others know about the specialty in the initial stage of its formation, the major research areas or problems that the communities of that particular specialty are focussing upon (Hoddeson et al., 1992). In this process it attracts new researchers to the group and also provides them some of the possible research problems.

Concerns of the Specialty in Teaching

Teaching is the medium of propagation of a subject. Through teaching the teacher enlightens his/her students. Thus, teaching and learning are interrelated aspects of a never-ending process. This is also a basis for social reproduction of the specialty. The content of teaching includes major achievements, both theoretical and experimental and experimental techniques.

Demographic Factor (Critical Size)

A nucleus or a core group creates the conditions for a specialty to sustain research for which a critical number of scientists is required. And this critical size is not fixed universally. It differs from specialty to specialty. Sub-critical size of scientists affects: (a) communication, (b) peer review system, and (c) the extent of impact that a specialty can make. If scientists working in a particular area do not communicate with each other, then their research would remain in isolation. A specialty to develop needs a critical number of scientists working in that particular research area (Mullins, 1972).

Social Reproduction

Social reproduction involves replenishment of human resources through recruitment to carry forward the research agenda of a specialty and the associated physical resources that have to be mobilised. Further, it involves production, dissemination of knowledge, sustenance and continuation of organisational units or a department in which the scientists are located, and is essential for the success of a specialty (Jagtenberg, 1983). Mentorship may be vital for continuity of thought, if not membership.

Leadership

Leadership is not a solitary activity. Sometimes, many people are involved in a group and the whole group leads the specialty. But again a leader may project his/her research areas/interests as the most important research areas/interests. It may result in a situation where the leader defines the field. If we take into consideration the model of Max Weber, then we will see that authority is of three types. We can use the word authority and leader synonymously. According to Weber (1905), these authorities are of traditional, charismatic and rational-legal types. The first type of leadership depends on the leader delivering a traditional message or holding a traditionally sanctioned office. The second type of leadership is focused on a certain quality of an individual personality by virtue of which she/he is set apart from ordinary women/men and treated as endowed with supernatural, superhuman, or at least specifically exceptional powers or qualities. And the third type is based on some generally agreed set of rules and procedures. In the book titled *Science in Action*, Latour (1987) describes the role of a leader in different stages of the development of a specialty. The roles the leaders play, is manifold. They have their own charisma to lead the discipline, they mobilise and allocate resources and work as an entrepreneur and contribute to the society by social reproduction, that is, by mentoring scientists capable of pursuing the same kind of research in future to enhance the productivity in that specialty.

We would like to examine the institutionalisation and development of solid state chemistry in India, by drawing upon the sociological accounts of specialty formation and its dynamics discussed above.

Solid State Chemistry as a Scientific Specialty

The discipline of solid state chemistry is an interdisciplinary scientific specialty which, intersects with some other disciplines like the physics, chemistry, engineering, materials science, solid state physics, metallurgy, etc.; solid state chemistry borrows tools, concepts and techniques from the above said scientific disciplines. During the last three or four decades the field of solid state science or the materials science has become increasingly interdisciplinary. Physicists and metallurgists have been joined by chemists in solving many problems relating to computer and communication technology, micro-circuitry, laser, photography and so on. During the early 1940s the studies carried out by Roland Ward and his co-workers in US emphasised the synthesis of new materials and the determination of their structure as the main focus of solid state chemistry. Since then growth in the field of solid state chemistry has been rapid and has undergone considerable change. The discovery of magnetic ferrites, semiconductors, superconductors, phosphors, cracking catalysts, etc., stimulated the search for new materials.

Mapping the Solid State Chemistry Specialty

Simultaneously, ceramists and metallurgists developed new methods for the processing of materials into useful components. The

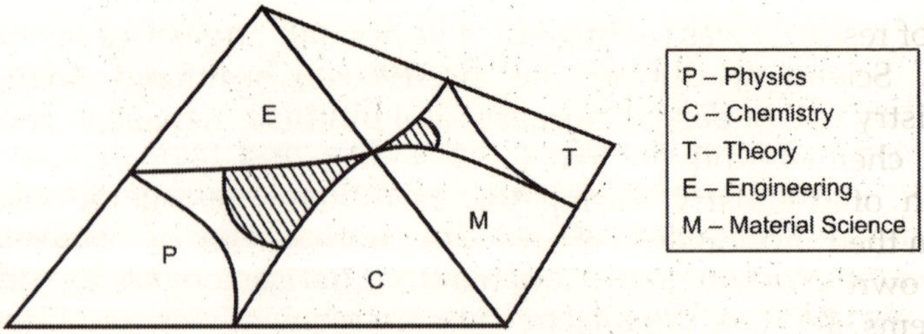

(Goodenough and Rao, 1983)

blurring of disciplinary boundaries took place as a result of the creation of interdisciplinary groups like materials research centres, which were needed to prepare and characterise more complex materials as well as to understand the relationship among synthesis, structure and properties. For this purpose the contributions of the chemists were essential. The solid state chemists can design, synthesize and characterise new materials. And they can also change, modify or create a new material by employing their expertise. Attempts to produce tailor-made materials of desired properties are an integral part of modern solid state chemistry. Some of the important sub-areas of solid state chemistry are Synthesis, structure and characterisation of properties of materials, Phase transitions, Electrochemistry, Glass science, Spectroscopy, Crystallography, Theoretical solid state chemistry, etc. During the 1970s the development of various techniques like X-ray Photoelectron Spectroscopy (XPES), Ultra Violet Photoelectron Spectroscopy (UVPES), etc., facilitated the research in the area of solid state chemistry in a more effective way. The solid state chemists also tried to improve the quality of the material which was at the same time less expensive and can have various applications.

As Holton (1972) notes, specialties arise out of scientists' intense competitiveness and interest in making new discoveries. The development of solid state chemistry took an interesting turn with the discovery of superconductivity phenomenon and subsequent research in the area of superconductivity during the late 1980s and early 1990s. Scientists worldwide worked on some or the other aspect of superconductivity. It was also recognised that solid state chemistry has potential industrial applications that can be used for the betterment of the society at large. The publication of the journal called *Journal of Solid State Chemistry* in 1969 made solid state chemistry visible as a scientific specialty.

Solid State Chemistry Research in India

Today solid state chemistry is a frontier area of the mainstream chemistry research in India. However, even a few decades ago, this field of research was pursued by a very small number of scientists in India. Scientists, during the pre-1960s, considered solid state chemistry to be a part of inorganic and physical chemistry. The traditional chemists did not recognise solid state chemistry as a new branch of chemistry during the 1960s. Perhaps all the scientists within the traditional chemistry departments were comfortable with their own research areas and never wanted to address the new problems, new phenomena, etc. The scientists who wanted to pursue research in the area specifically concerning solid state chemistry had to separate themselves from the traditional chemistry departments. One of the major motivations of the chemists behind the shift from inorganic and physical chemistry to solid state chemistry was driven by the need to make tailor-made materials. This motivation was influenced by the condensed matter physics research where the aim was to make newer and better transistors which were used in making defense related materials. Subsequently, this special branch of chemistry gradually became prominent and developed its own research programmes.

As mentioned above, till the early 1960s, the small number of scientists engaged in solid state chemistry research were scattered throughout the universities, National Laboratories, Institutes of Technologies, etc. But it was C.N.R. Rao, one of the pioneering solid state chemistry researchers of India, who played a significant role in giving impetus and direction to solid state chemistry research. Solid state chemistry research was rigorously pursued by him and others after the establishment of the SSCU. Before the establishment of the SSCU he was at IIT (Indian Institute of Technology), Kanpur. He, along with other scientists, conducted several winter and summer schools, devoted to introducing solid state chemistry, at IIT, Kanpur. After the establishment of the SSCU at the IISc, he continued to conduct winter schools, meetings, seminars, etc., with the help of his colleagues. He also invited many eminent scientists working in the area of solid state chemistry, from abroad, to these meetings. The chief motivation behind conducting these schools, meetings, etc., was to motivate scientists from other universities and institutions of the country to recognise the tremendous scope of solid state chemistry and initiate them into solid state chemistry research with a view to form a scientific community of solid state chemists in India. Gradually, many scientists recognised the challenging opportunities solid state chemistry might provide, for illustrating and applying principles of solid state chemistry to systems of theoretical interest and technological importance, and took up research in the area of solid state chemistry. Hence, it is clear that solid state chemistry was institutionalised as a specialty recently in India due to the vision of the pioneer researcher.

Establishment and Successful Consolidation of the SSCU

In November 1976, the Solid State and Structural Chemistry Unit (SSCU) was formally established at the Indian Institute of Science (IISc), Bangalore, by C.N.R. Rao. He visualised the importance of the subject and established the Unit, which is fully devoted to solid state chemistry. As he said, the motivation behind establishing a department totally devoted to solid state chemistry, which would enable scientists to focus on a set of related problems in a given specialty, was:

> Every subject needs a group of people working in it. In a department there may be ten people working in ten different areas. It may be good for the department, but not good for the subject. More effective groups are small groups. One should give freedom and autonomy to these groups.

The above statement given by the founder-leader of the SSCU corroborates the observations of Price (1963) regarding the emergence of a large number of smaller specialties. According to Price, scientists' research activity tends to be highly specialised and major changes in a variety of disciplines have been generated within small, socially coherent groups. They can absorb technical information only from a limited number of sources. Although the scientists of the SSCU were from different disciplinary backgrounds, they could communicate with each other and were concerned with problems closely similar. As the scientists of the SSCU were smaller in number and socially coherent they could consciously form a group which was fully devoted to the teaching and research of the specialty called solid state chemistry in India.

Rao was able to mobilise resources and expertise to establish the SSCU. The leader already had noticeable intellectual accomplishments and hence had achieved higher recognition than any of his contemporaries. He could pursue research successfully in different sub-areas of solid state chemistry which was on par with the international standards. And this is evident from the number of his publication in reputed international journals. He is a Fellow of Royal Society of London and received many awards (nearly 25 awards both of national and international repute during the study period under consideration). These awards area measures of the recognitions he achieved. His leadership did contribute to the success of the SSCU and to an extent to the spread of solid state chemistry research in India. Many young students were attracted by his intellectual leadership and joined the Unit to pursue their research. The Unit started with only one scientist, that is, the founder himself in 1976.

Later many scientists, who specialised in some area of solid state chemistry, were invited by the leader to the Unit gradually. Though the leader had interest only in a few the sub-areas of solid state chemistry, he meticulously chose the scientists having expertise in different sub-areas of solid state chemistry and appointed them to give the SSCU the status of a department totally devoted to the teaching and research of solid state chemistry. The scientists who joined in course of time gradually developed their individual research programmes and also participated in collaborative research work with other scientists. They shared resources available in the SSCU and helped each other by sharing individual expertise they have, to solve a problem.

Besides the visionary leadership, the other factors which contributed for the growth of this Unit were the physical and human resources. The founder-leader played a crucial role in creating the infrastructural facilities that were of international quality to carry out solid state chemistry research in a very effective way within India. The Unit located in one of the premiere science research institutions in India had some initial advantages. The kind of research funding that this Unit received were more than any other research centres, located in different institutions of India where solid state chemistry research was carried out,. This Unit attracted very bright doctoral students and postdoctoral fellows. The number of doctoral students has been increasing every year. The scientists who were pursuing their respective research in the SSCU were also reputed both nationally and internationally.

Twenty-Five Years of the SSCU: Research Output and Visibility

As has already been mentioned in the beginning, this study took into consideration twenty-five years of the establishment of the SSCU. Here we would like to provide a brief account of the research output and visibility of the Unit. Besides developing its own research and teaching programmes in solid state chemistry, the Unit has strong interaction with other departments of the Institute. Interdisciplinary and interdepartmental research programmes form a significant part of the Unit's activities. Thus, the Unit offers some courses which were open to some departments of the Institute and some other institutes nearby. The Unit organised many seminars, conferences and winter schools and invited many eminent speakers from various parts of the country as well as from outside the country. During our study, that is, in 2004, there were ten working scientists in the Unit. The scientists of the SSCU were from various educational backgrounds. The scientists were also expert members of many advisory committees. The scientists at the SSCU secured research grants from various sources, both governmental and non-governmental, to carry out their research. The number of research projects (both new and on-going) that this Unit got vary from year to year over a twenty-five year time period (1976–2001) the minimum being three and the maximum being sixty three. Starting from the establishment of the Unit in 1976 till 2001, the Unit has produced nearly 2000 research papers in different areas of solid-state chemistry, spreading across nearly 200 different journals published worldwide. Apart from the scientific articles, the scientists have written many books, which are considered as the basic books for those who want to pursue research in the field of solid state chemistry. For example, the founder leader of the SSCU has written/edited more than 35 books. And probably the first systematic book in this field of solid state chemistry named *'New Directions of Solid state Chemistry'* was brought out by two faculty members of the SSCU. The scientists at the SSCU have several patents. A single scientist has four Indian patents. Another scientist has three U.S. patents. Three of the faculty members have ten patents altogether. As the SSCU is basically a research-oriented one, the Unit admits students for PhD programme. The number of PhD students varies from year to year. But the trend shows an increasing number of students coming and joining the unit, the minimum being 10 (1977) and the maximum being 58 (2001) keeping the study period from 1976 to 2001.

Success of the Unit

The scientists of the SSCU received many awards and were invited to occupy advisory positions of many institutions and funding agencies, membership of editorial boards of journals. The alumni of the SSCU were well placed in different institutions within India as well as outside India. Other eminent scientists of India acknowledge that the SSCU is a highly visible department totally devoted to the research and teaching of solid state chemistry in India. People from outside the country also recognise this department as one of the best solid state chemistry departments in the world. As the founder-leader put it, "now in the whole world the SSCU is a well-known chemistry department".

The social reproduction, as mentioned above, of the solid state chemistry community is facilitated by the SSCU as it has formal teaching and research programmes to train the young scientists to pursue their research career in solid state chemistry. Scientists were appointed in this Unit to train students as well as pursue their own individual research. Besides developing its own research and teaching programmes, the Unit has strong interaction with other departments of the Institute. Interdisciplinary and interdepartmental research programmes form a significant part of the Unit's activities. Although the scientists of the SSCU were from different disciplinary backgrounds, they could communicate with each other and were concerned with problems closely similar.

Besides interacting with each other within the SSCU, the scientists of the SSCU also interact with scientists in different departments within the IISc. As solid state chemistry is interdisciplinary in nature, it draws upon both tools and concepts from different disciplines. The SSCU also offers some interdisciplinary courses that are open to all the other relevant departments of the IISc. On one hand, the individual departments within the IISc, like the SSCU, help in augmenting the visibility of the IISc the national level as a premier science institute. On the other hand, the IISc patronises these individual departments as organisational units and provides these departments various resources to carry out research.

Collaboration of the SSCU Scientists

Collaboration as a concept means working together. The constituent elements of the process of collaboration are contract, interaction and communication (Jeffrey, 2003). The strength/expertise of collaborating individuals creates conditions for interaction on the basis of equality. There are various types of collaboration. Collaboration can be done in terms of institutional affiliation, academic affiliation or industrial affiliation. The present study took into consideration the following types of collaborative work in terms of the SSCU scientists having: (a) collaborative work with their own laboratory group (Own); (b) with the founder-leader (K1*); (c) among themselves (K1); (d) with scientists of different departments within the IISc (K2); (e) with scientists from different institutions of India outside IISc (K3); and (f) with scientists from other countries (I).

Table 1

Approximate Percentage of Collaborative Publications of each Scientist for the Period between 1976 and 2001

Scientist #	*Own*	*K1**	*K1*	*K2*	*K3*	*I*
Scientist 1	39	—	26	17	5	12
Scientist 2	29	47	8	9	1	4
Scientist 3	69	17	3	3	—	8
Scientist 4	38	16	10	0.5	2	35
Scientist 5	20	17	11	32	14	6
Scientist 6	53	1	16	7	4	19
Scientist 7	50	3	4	15	—	28
Scientist 8	49	16	6	6	19	4
Scientist 9	33	12	8	4	1	42
Scientist 10	78	14	—	2	6	—

From Table 1, one would observe that the greater the extent of collaboration of the scientists of the SSCU with the leader, the smaller is the extent of collaboration with scientists from outside the SSCU (K2, K3, I) and among themselves (K1).The scientists whose collaboration with the leader is not as pronounced as in the case of the other scientists seemed to have greater collaborative research with their own groups and with scientists outside the SSCU including those in the foreign countries. Most of the scientists had a greater degree of collaboration with his research group. The collaborations were mainly based on complementarity of expertise. The scientists appointed by the leader continued to work in the research areas for which they were recruited, thus, actualising the vision of the leader. In other words, along with their attempts to realise the vision of the leader, the scientists of the SSCU also pursued their individual research. But after the leader retired in 1994, scientists at the SSCU pursued their individual research in a much more prominent way. The effectiveness of collaboration of the scientists within the SSCU was very negligible compared to other forms of collaborations. Scientist 10 did not have any collaboration at all with the scientists of the SSCU. Scientist 5 specialises in instrumentation and scientist 8 had expertise in crystallography. Some of the scientists of the SSCU had collaborative work with these two scientists mainly for the expertise that these two scientists have. There was only a single strong collaboration between scientist 4 and scientist 6. They had collaborative work in the area of proton conduction and ionic conduction. As scientist 4 said:

> Collaboration is based on equal interest in the problem and equal strength. If there is a problem, I can do something and there is another scientist who can bring another dimension to that problem. That kind of successful professional collaboration I have here with (scientist 6). He is an electrochemist and has expertise to measure certain electrochemical properties of solids. I am a synthesist and I don't have expertise to measure those properties. So I both came together.

So, the intra-SSCU collaboration of the scientists was complementary in nature. One may say that the SSCU scientists developed an organic solidarity among themselves as suggested by Law (1973). In this kind of division of labour, scientists come into relationship with one another because one performs services which, the other can't easily carry out for oneself (Law, 1973). The above interview excerpt from the scientist 4 corroborates the thesis of organic solidarity.

Since solid state chemistry is interdisciplinary in nature, the scientists of the SSCU had collaborative work with scientists from various departments within the IISc. If we compare the percentage of collaboration of the SSCU scientists with the scientists from various institutions of India, other than the IISc, with that of the scientists from other countries, then we find that the percentage of collaboration with scientists from other countries was greater than the collaboration with the scientists in India. This further indicates that the SSCU scientists did not have much collaborative work with scientists of their own country and had greater collaboration with the scientists in other countries. During our study the SSCU scientists were asked about the insignificant collaborative work with their peers outside the institute within India. Some of the reasons for the insignificant collaboration, according to the SSCU scientists were, "Indian scientists delay doing things", "the collaboration was not on equal basis", "there was not much research work on solid state chemistry area in other parts of the country", "lack of motivation among Indian scientists", etc.

As scientist 4 responded:

> I tried to collaborate with some of the scientists in India. But somehow the collaboration with international people is much easier and faster. Once they accept something, they go forward. That is their culture. But in India, people delay doing things because our culture is like that. But I have very successful collaboration with another scientist here. We solved some problems which we couldn't have done individually.

Scientist 5 specialising in the area of spectroscopy gave another kind of response as:

> In India collaboration is a badly treated word. All collaborations are badly done, because every scientist is underestimating the other fellow. This is the Indian crab syndrome. So in India much trust and much companionship are needed.

However, the responses from solid state chemists located in other institutions of India don't corroborate with the perceptions of the SSCU scientists. The other scientists felt that due to lack of funding they were unable to do the kind of research they wanted to do. They are also capable of carrying out research in frontier areas of solid state chemistry if proper infrastructural facilities are provided. They blamed the SSCU for getting disproportionately higher share of funding.

The SSCU: Centre of the Periphery – Periphery of the Centre

The concepts of 'centre' and 'periphery' have been in use in the sociological literature to describe the economic exploitation of one country by the other. But while borrowing these concepts of *centre* and *periphery*, the present paper has given a different content to the meaning of the two related concepts. The components of the content relevant for the present study are: (a) setting research agenda; (b) choice/selecting experiment and experimental techniques; (c) identifying research problems; and (d) potential clientele for the research output, etc. In other words, we can use the concepts of 'centre' and 'periphery' to describe the development of solid state chemistry in India in terms of the research agenda, research concerns, critical size and pattern of collaboration of the scientists at the SSCU. The status of the SSCU in terms of its interactions or linkages is two-fold. On the one hand, it is linked to the International Solid State Chemistry community and on the other hand it is linked to the members of the solid state chemistry community located in various institutions within India. We may argue that the SSCU should be considered as the *periphery* in relation to the centres of solid state chemistry research in the west and as the *centre* in relation to the solid state chemistry research groups located in various institutions, in India.

Let us look at the reasons for considering the SSCU as a periphery in relation to the centres of solid state chemistry research in the west. This can be seen from the point of view of the research areas that were undertaken by the scientists at the SSCU. In the present study it was found that the scientists of the SSCU were taking up the research areas that were pursued by the scientists of the western countries. This can further be established by looking into the publication of articles in different international as well as national journals by the SSCU scientists. Most of the research output of the SSCU scientists was/is published in the international journals itself indicates that the SSCU scientists were pursuing research areas that were seen as significant and relevant by solid state chemistry researchers in the west. We noted that research agenda – both, basic and applied – relevant to the Indian context were not considered worthwhile agenda. Graph 1 indicates that out of nearly 1,800 articles published by the scientists of the SSCU, 1,617 were in various international journals and only 185 were in different national journals. And publishing in international journals also indicates that the scientists themselves were assigning priority to the international journals.

Here, there is again a little problem and I would like to discuss about the very concept and definition of an international journal. What we consider as an international journal like *Physical Review B*, *Journal of Solid State Chemistry*, *Chemistry of Materials*, etc., are actually journals published from their respective countries. What

Graph 1

Publication of SSCU scientists in National and International journals

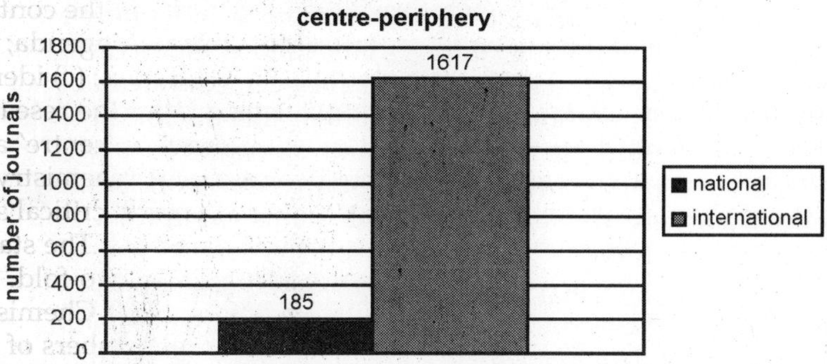

makes these journals international is the contribution from the other countries. Most of the good researches from most of the countries are getting published in these so-called 'international journals'.

Lack of infrastructural facilities in India compared to the western world provides us some reasons to consider the SSCU as a *periphery*. Another factor which can add to the argument is in terms of the critical number of researchers working in the area of solid state chemistry. In the west the number of solid state chemistry researchers is much higher than the number of solid state chemists in India. It is always not possible to demarcate between the centre and periphery in this manner. Within the western world itself, there exists a centre-periphery relationship as reflected by the more prestigious institutions and the less prestigious institutions. Future research may provide us somewhat more concrete reasons to establish the hypothesis that SSCU is a periphery in relation to the western world of solid state chemistry.

In relation to the solid state chemistry centres within the country, the SSCU may be seen as the *centre* and other academic institutions practising solid state chemistry may be considered as the *periphery*. This is indicated by some pertinent points. The location of the SSCU within one of the most prestigious institutions of India, that is, IISc is one such point. The kind of funding and students that the SSCU gets is qualitatively higher and better than most of the other centres located in various institutions of India, where solid state chemistry research is carried out. The resources and infrastructural facilities that are available in the SSCU are the best in India to pursue solid state chemistry research effectively. As Mendelsohn et al. (1977) argued the type and location of employment becomes a major factor in stratifying scientists and available resources in professionalised sciences. Indeed, scientists at low prestige institutions like the peripheral institutes pursuing their research in the area of solid state chemistry in India may not even conceive of undertaking daring innovations or seeking major rewards. By contrast, scientists who have been educated at scientifically and socially authoritative institutions like the IISc or the SSCU consider it natural to make inspired guesses, to attack major problems and obtain and organise resources for work in new areas, especially if they are also from a high status stratum in solid state chemistry community.

Another point that indicates the centre periphery relationship between the SSCU and other institutes of India pursuing their research in solid state chemistry area is in terms of membership in various organisations. The ISCAS (Indian Association of Solid State Chemistry and Allied Scientists, established in 2001) was formed by the scientists, especially those located in other centres or institutions, who felt that their research concerns were not addressed properly by the existing societies. In fact, most of the practising scientists of the SSCU were either unaware of the ISCAS or were not interested in becoming members of that society. And the societies like MRSI (Material Research Society of India, established in 1989) were dominated by the scientists from a few prestigious institutions like the IISc of India.

The Solid Community: The Role Played by the SSCU in India

The SSCU, as a Unit devoted to solid state chemistry teaching and research, has not been replicated elsewhere in India. The SSCU scientists have also not invested adequate time and energy towards the spread of solid state chemistry as a specialty to other institutions in the country. This may be the reason as to why the specialty still lacks the critical number of scientists working in this area. Another factor that contributed to the sub-critical size of the solid state chemists' community is the 'Brain Drain'. Many of the good scientists trained in solid state chemistry have been migrating to other countries, especially to the US and Europe, due to better job opportunities. The SSCU also contributed to the brain drain. Several students who did their doctoral work at the SSCU are currently in the US or Europe. The solid state chemists of the SSCU did not have collaborative work with the scientists of the other institutions of India as mentioned above. Even though the scientists of the SSCU were in the *centre*, they have not done much, in terms of providing research careers, challenging research problems to check the brain-drain. Conditions should be created inside the country so that they would prefer to stay in the country rather than emigrating to other countries. Industries and service sectors that need inputs from solid state chemistry should come up to provide research career opportunities in R&D for solid state chemistry researchers.

It should be acknowledged that, whatever popularity solid state chemistry as a specialty got in India is largely due to the efforts of the scientists of the SSCU and especially the founder-leader. Though this Unit may not have served as a role model to other institutions, scientists of the other institutions became aware of the research contributions of the scientists at the SSCU.

Conclusion

As an organisational unit, the SSCU appears to be unique, owing its establishment to the vision of the founder-leader. Leadership is one of the prominent factors in establishing the SSCU. Since solid state chemistry is interdisciplinary in nature, the scientists of the SSCU were interacting with scientists from other departments to solve problems. However, the SSCU scientists did not have much collaborative work with the Indian scientists outside the IISc compared to the collaborative work with scientists of the other countries. The SSCU as an organisational model has not been replicated elsewhere in India. The scientists of the SSCU, being in the centre of the solid state chemistry research, did not spend adequate time and energy to propagate solid state chemistry widely throughout India. But it will remain a fact that whatever popularity solid state chemistry as a specialty got in India is due to the efforts of the scientists of the SSCU. The solid state chemistry researchers in India haven't attained the critical size so far to build a solid community of solid state chemists in India.

Due to various promising potential application of solid state chemistry, the industry has been playing an important role in the development and progress of the solid state chemistry specialty in the developed countries. This may be one of the important reasons as to why solid state chemistry is flourishing in the developed countries. But In India, solid state chemistry, as a scientific specialty, insulated itself from the industry and depended almost entirely on government funding for sustaining their research. If the scientists of the SSCU would have taken interest in solving problems that were specific to Indian industry, some organic linkages would have been established with society at large through the industry. The growth of this specialty would have taken a different shape in India. It will, therefore, be interesting to examine the specificities as the one mentioned above regarding lack of interaction with industry, if any, of the larger social milieu which contributed towards the specific pattern of growth and sustenance of this specialty in India.

Note

All the names of the scientists are not disclosed in this paper. Whenever excerpts from interviews are mentioned, the names of the scientists are withheld to maintain confidentiality and each scientist is identified by a number.

References

Beardsley, Edward H. (1964). *The Rise of American Chemistry Profession, 1850–1900*. Florida: University of Florida Press.

Ben-David, J. (1984). *Scientists' Role in Society: A Comparative Study*. Chicago and London: University of Chicago Press.

Bitz, A., McAlpine Andrew, Whitley R.D. (1975). The Production, Flow and Use of Information in Research Laboratories in Different Science.

Bloor, David. (1976). *Knowledge and Social Imagery*, Chicago: The University of Chicago Press.

Brock, W.H. (1992). *The Fontana History of Chemistry*. London: Fontana Press.

Cetina, Karin Knorr (1981). *The Manufacture of Knowledge: An Essay on the Constructivist and Contextual Nature of Science*, London: Pergamon Press.

Cole, Jonathan and Harriet Zuckerman. (1975). "The Emergence of Specialty: The Self-Exemplifying Case of the Sociology of Science", in L.S. Coser (ed.), *The Idea of Social Structure: Papers in Honor of Robert K. Merton*. New York: Horace Brace Jovanovich, pp. 139–74.

Crane, Diana. (1972). *Invisible Colleges: Diffusion of Scientific Knowledge*. Chicago and London: Oxford University Press.

Ellis, A.B. et al. (1993). *Teaching General Chemistry: A Materials Science Companion*. Washington, DC: American Chemical Society Publication.

Haribabu, E. (1999). "Scientific Knowledge in India: From Public Resources to Intellectual Property", *Sociological Bulletin* 48(1 and 2): 217–33.

Hoddeson, Lillian, Earnest Braun, Jurgen Teichmann and Weart Spencer (eds.) (1992). *Out of the Crystal Maze: Chapters from the History of Solid-State Physics*. New York: Oxford University Press.

Jagtenberg, Tom. (1983). The Social Construction of Science, Holland: D. Riedel Publishing Company.

Jeffrey, P. (2003). "Smoothing the Waters: Observations on the Process of Cross-Disciplinary Research Collaboration", *Social Studies of Science* 33(4).

Joshi, S.K. and R.A. Marshelkar. (eds.) (1995). *Solid State Chemistry: Selected Papers of C.N.R. Rao*. Singapore: World Scientific Publishing Co. Pvt. Ltd.

Kadushin, A. (1968). Games People Play in Supervision. *Social Work*, 13, 23–32.

Kuhn, T.S. (1964). *The Structure of Scientific Revolutions*. USA: University of Chicago Press.

Latour, Bruno and Steve Woolgar (1979). *Laboratory Life: The Construction of Scientific Facts*, Princeton: Princeton University Press.

Latour, B. (1987). *Science in Action: How to Follow Scientists and Engineers through Society*. USA: Harvard University Press.

Law, John. (1973). "Development of Specialties in Science: The Case of X-Ray Protein Crystallography", *Science Studies* 3.

Mendelsohn, Everett, P. Weingart and R. Whitley (Eds.). (1977). *The Social Production of Scientific Knowledge: Sociology of the Science*, Dordrecht: D. Riedel Publishing Company.

Merton, R.K. (1973). *The Sociology of Science*. London: Chicago University Press.

Mulkay, Michael, Milic Vojin and Starr M. Jerold. (1980). *The Sociology of Science in East and West*, Sage Publication.

Mulkay, M.J. (1980). "The Sociology of Science in East and West", *Current Sociology* 28: 1–181

Mullins, Nicholas C. (1972). "The Development of a Scientific Specialty: The Phage Group and the Origin of Molecular Biology", *Minerva* 10: 51–63.

Price, D.J. de Sola. (1963). *Little Science Big Science*. New York: Columbia University Press.

Raina, Dhruv and B.M. Gupta. (1998). "Four Aspects of the institutionalisation of Physics Research in India (1900–1950): Substantiating the Claims of Historical Sociology through Bibliometrics", *Scientometrics* 42(1): 17–40.

Rao, C.N.R. (ed.) (1974). *Solid-State Chemistry*. USA: Marcel Dekker, Inc.

Rao, C.N.R. and J. Gopalakrishnan. (1987). *New Directions in Solid-State Chemistry*. Great Britain: Cambridge University Press.

Restivo, S. and Bauchspies. (1996). W. *Sci. Cult.*, 6: 396–413.

The Auld Lang Syne of the South Carolina Conference. An Address Delivered before the Historical Society of the South Carolina Conference, Methodist Episcopal Church, South, at Spartanburg, South Carolina, December, 12, 1905.

Weber, Max (1905). *The Protestant Ethic and the Spirit of Capitalism*. London: Routledge and Kegan Paul.

Whitley, Richard (ed.) (1974). *The Social Process of Scientific Development*. London: Routledge and Kegan Paul.

Zuckerman, H. (1988) "The Sociology of Science", in N. Smelser ed., *Handbook of Sociology*, Newbury Park, CA: Sage, pp. 511–574.

20 Technology of Defence Production in India under British Rule

Cossipore Factory in the Late Nineteenth and Early Twentieth Centuries

Arun Bandopadhyay

Perhaps the single most important feature of modern warfare was the domination by modern weapons. The industrial underpinnings of the nineteenth century revolution created a massive change in the organisation and technology of war, which were then quickly mastered by the European countries on a highly competitive basis. Tools or weapons were made in close connection with modern science and technology, and if only the right ones could be discovered they could ensure victory to a great extent. "Extended warfare between the great powers intensified efforts to apply scientific knowledge and inventiveness to the business of destruction."[1] Suddenly, high superiority of weapons became the desired objective of all, and other aspects of military preparation turned into things of secondary significance.

Naturally, revolutionary changes set in the technology of defence production, particularly in small arms and artillery, from about the middle of the nineteenth century.[2] Between 1815 and 1870, the weapons of both infantry and artillery in Europe were transformed. The muskets of the Napoleonic era were smoothbores, accurate at 50 yards and the maximum range was 200 yards. The guns were also smoothbore muzzleloaders, as all the guns had been since the fifteenth century. "But by 1870, the principal armies of Europe were equipped with rifles accurate up to 500 yards and with rifled guns with ranges between 2,000–3,000 yards."[3] Between 1875 and 1914, the consumption of firearms and artillery increased enormously. The British field artillery was inferior to that of the

European Powers in range and rapidity of fire.[4] In 1889, the field artillery of the French and German armies could have barrels that alone recoiled against a progressive buffer action instead of the whole carriages. "This made possible a faster rate of fire and used shrapnel shells which burst in the air above the enemy."[5]

Under the conditions of general competitiveness among the European countries, the artillery and the engineering branches of the armed forces were the first to acquire a professional outlook, thus linking military technology to the rise of professionalism. This outlook spread from one country to another and by the end of the nineteenth century, a loose term called the Military-Industrial Complexes came into being, "comprising the totality of the institutions involved with developing, manufacturing and manning the hardware of war".[6] The very requirement of war in such a changing scenario presupposed a state-of-the-art technology in the production of arms and ammunitions, failing which there was always a risk of the product becoming obsolete and, therefore, useless. "In war, especially in modern wars, wars in which weapons change(d) rapidly, one thing was certain; no army of 50 years before any date selected would stand 'a day's chance' against the army existing at that date."[7]

Herein lay a basic dichotomy between the technology of defence production and the general dependence of that production as promoted by colonialism. Technology did not permit backwardness and laxity in this world of rapid changes but colonialism often failed to rise up to the task, at the right time and in sufficient measure. There was an inherent limitation in colonialism for the development of a really independent, modern and dynamic technology of ordnance production. The very nature of the contradiction usually led it to sacrifice the interests of the colonised country for the sake of the coloniser. However, the paradox of the situation was that the technology of defence production, as understood in modern times, had also an inherent tendency to sometimes get out of the limitations generated by colonialism for its own sake. This dichotomy and deviation, pursued so dramatically in British India, form the main theme of this chapter.

Cossipore under Three Distinct Phases of Technology

The early years of Cossipore's technological experience were very promising, particularly after the establishment of the Gun Foundry there in 1830. The most notable characteristic of Captain Hutchinson's experiments of the 1830s was the display of high engineering feats by the contemporary world standards. There is also reason to believe that production technology did not lag much behind. We have seen certain degree of optimism during this period, so much so that with the introduction of the cupola furnace by the middle of the 1830s, Hutchinson could write that Cossipore would require no more cast iron articles in future for this purpose. This phase of promise and hope continued till the onset of the Mutiny when Captain Broome made some notable additions and experiments and Cossipore, "beside producing brass ordnance, turned out a very large stock of iron work of all descriptions including machinery and carriages of a Light Field Battery".[8] It is probable that Cossipore was given a relative degree of freehand in its production, which was automatically reflected in the use of its technology.

For about thirty years since 1858, Cossipore's technological experience was discouraging. There was obstruction rooted in the change of imperial policy of Ordnance in India. Cossipore was not given the sophisticated machinery available at that period of technological change. The mode of construction of even the brass Ordnance became the copy work of the mode practised in Woolwich. It was even stated that the Ordnance Select Committee at Woolwich would decide what Cossipore should produce. The respective Governor Generals of India from Canning to Northbrook did not try to check the process, despite appeals from influential quarters. The latter even paved the process by his decision of virtual stoppage of gun-making in Cossipre in July 1873. However, the ongoing projects of the manufacture of 9-pounder muzzle-loading rifled projectiles could not be immediately closed and Northbrook had to write to the Secretary of State for India in August 1873 for the import of "such quantities of (pig) iron, in addition to that to be locally purchased and under supply, as may be deemed sufficient after consideration of Colonel Maxwell's statement" for the work.[9] Known as Foundry and Steel Factory (FSF) at that time, Cossipore of course tried its best to get out of the compulsions imposed from above. For example, as early as 1875, Major T. Nichol, the Superintendent (1874–1882) of the of the factory complained to the Inspector General of Ordnance and Magazine that the steam drop in the Shell Factory was only capable of forming bevels of the shrapnell shells up to 80 pounders but was entirely occupied with the 9-pounder projectiles. For the extra production of 9 x 7 pounder shells, a second steam drop similar to the one in use in the Royal Laboratory, Woolwich was necessary. He believed that Cossipore on its receipt would be able to turn out shrapnell shells for the 7-inch, 8-inch, 9-inch and 10-inch guns and also for Palliser shells.[10] Such complaints were not seriously considered and there was no immediate reflection in Cossipore's production. The FSF was only able to project considerable re-direction of its operation and technology by the early 1880s.

The requirement of the new items of defence production along with their changing technology in the last quarter of the nineteenth century made considerable inroads on the policy of colonial dependence of Ordnance production in India. The cost factor of production was another cause for the modification of the imperial policy towards Cossipore. Cossipore was given more active assistance in terms of equipments and personnel henceforward. Consequently, a process of relative 'freehand' in technological matters was initiated, first in the 1890s, then in the 1900s and finally after the formation of the Munitions Board during the First World War. However, the change of policy was hinted as early as November 1887, when India Office wrote on behalf of the Secretary of State to the Governor General of India, requesting the latter to supply information about the plants to be required by the Cossipore Factory for the manufacture of French time fuzes by the latest technology. Of course, more important changes were in the offing, and in the next 20 years Cossipore's technological experiments in the production of steel (in the 1890s) and the manufacture of field guns (in the 1900s) would make a major reshuffle of preference and priorities in ordnance production in India under colonial dependence. From 1905 onwards, Cossipore was known as Gun and Shell Factory (GSF). We shall discuss these two experiments in some details in the following two sections.

New Technology and the Manufacture of Steel in Cossipore

In the history of modern iron and steel production in India, the success story of Tata Iron and Steel Company (TISCO) in the early years of the twentieth century is well-known and widely appreciated.[11] This is quite legitimate if we consider the range and scale of operations and the developments that were followed in this unit of production at what is now known as Jamshedpur. At TISCO, the actual production started in 1911–12.[12] However, more than fifteen years before this historic beginning, steel was successfully produced in the FSF, Cossipore by modern technological methods. This story is least known.

Several serious attempts had been made to produce modern iron in India in the early nineteenth century, but they failed. Individuals who undertook these ventures were mostly lured by official support, the actual scale of which is unknown. The earliest attempt was made by a former Company servant named J.M. Heath who was given by the East India Company an exclusive right to produce iron through methods 'similar to those in Europe'. He "undertook production at Porto Novo, a small sea port south of Madras, at Palampetti near Salem, and at Beypore on the Malabar

Coast. Charcoal was used as fuel and bullocks were employed to power what machinery was used".[13] This experiment failed. What was more important, George Jessop of the Jessop and Company at Calcutta protested against this grant of exclusive right to Mr Heath in 1825. He identified himself as a partner in one of the largest iron foundries in Great Britain and submitted:

> The works belonging to myself and partners in Calcutta were purchased by my brother, Mr H. Jessop, for the purpose of forming a connection with the Butterly Iron Works at home. It is therefore not to be expected that under such circumstances I should wish to manufacture in this country an article the sale of which I am interested in promoting in my own country and which formed so great a revenue and gives employment to thousands.[14]

For various reasons, Jessop did not initiate the project, and half a century lapsed before a more serious effort was made in Bengal to produce steel under modern lines. With the development of railways and other public works and the rise of modern industry in India after mid-nineteenth century, India became a large importer of iron and steel, and plans were undertaken to consider the possibilities of import substitution. "The Bengal Iron Works Company, the first plant designed along modern lines, was formed in 1874 to take advantage of this growing demand and the high prices of imports."[15] Its first blast furnace and foundry came into operation in 1877, when it was overtaken by a financial and operational crisis. "It was discovered that the plant had been designed on too small a scale and along outmoded lines ... (and) the company shut down in 1879."[16] The Government of India purchased the defunct company at a nominal price and operated it as a public enterprise for some time with limited success. In 1889, doubts and pressures from Home Government and private entrepreneurs compelled the Government of India to finally close it down as a public enterprise.

It was at this juncture that the thought for the manufacture of steel under recent technological method was initiated in Cossipore. The Bessemer steel process, introduced in Europe in 1856 in response to the demand of Napoleon III for steel capable of withstanding new explosive shells, had made a breakthrough in defence production. "The Bessemer, open hearth, and Gilchrist processes of steel manufacture lowered the costs of production and led to the general use of steel, in place of iron, for heavy artillery in the last quarter of the nineteenth century."[17] Much of the failure in the previous attempts to make durable steel in India was rooted in technology. Naturally, years before the erection of the extensive iron and steel foundry in Cossipore in 1892, serious considerations were given for the appropriate technology of the manufacture of steel. As early as July 1887, Captain R.H. Mahon, then officiating Superintendent, FSF, Cossipore wrote to the Inspector General of Ordnance, Bengal:

> To manufacture either the cast or forged steel projectiles in India for so many different guns and to set up the large amount of machinery necessary will cause an immense outlay and it will finally commit the Government to a method of manufacture which may after all be superseded by the cheaper method introduced by Messrs Nordenfeldt and Noble ... I know of no reason why such a test should not be carried out in India ... .[18]

The method that Mahon was referring to was described in contemporary reports as Mitis castings. As E.A. Cowper, the late President of the Society of Mechanical Engineers in England explained, the surface of the Mitis casting was remarkably smooth owing to the special composition of the moulds and they required little cleaning upon an emery wheel. "The Mitis casting c(ould) be made in *Chill moulds* without becoming brittle or bad in any way and *never* require(d) annealing."[19] The advantage of casting 'chill' in shell work was considered to be very great, making a large reduction in cost by saving labour, time and material required for sand moulding The superiority of the Mitis iron as a casting metal was due to its greater purity.

The Mitis system was also useful in "other work than shell-making". It was claimed that it would furnish a method of doing away, not only with forging almost altogether, but also of greatly reducing the labour of working up the rough forging with machine and other tools. Captain Mahon continued:

> I have examined a number of castings sent out to this country by Major Wace, and one or two I have worked in machines. The castings are wonderfully clean and homogenous and are cast exceedingly light; this was especially noticeable in the case of a ploughshare which in its thickest part scarcely exceeds 3/8 inch and withal very strong. I find them work up without difficulty, though a trifle harder than ordinary castings.[20]

Mahon was not the only person to have such faith in the discovery of the Mitis process of steel manufacture in India. Colonel H.J.F. Shea (1886–1891), the Superintendent of the Gun Carriage Factory, Fategarh was also impressed by it. In his letter to the Deputy Inspector General of Ordnance, Bengal Circle, in July 1887, he identified a good number of items likely to be made under this system – carriages for horse, field and heavy batteries, framework of wagons, axletrees, naves, shift irons, limber hooks, trail eyes handles and hinges of ammunition, boxes, nuts, bolts, screws, rivets, stirrup irons, footrests etc. These items then "should be tested in England for (the) Government of India *even if authorities at Woolwich do not intend to entertain the idea of promptly proving the merits of Mitis castings* ...".[21] The Director-General of Ordnance in India was aware of the apparently most valuable discovery and its possible benefits, but as the system had not generally adopted in Europe he advised the Government of India in August 1887 "to proceed with caution before introducing it into India for the manufacture of various Ordnance stores".[22] Major R. Wace, who had been the Superintendent (1882–1890) of FSF, Cossipore but was deputed on special duty to England during this time, was more optimistic and emphatic. In his letter to the India Office, London in September 1887, he was eloquent on his recommendation.

I recommend that a supply of rough forging sufficient to meet all the requirements for the next two years be sent to Cossipore, as they are now supplied to Woolwich; and at the same time that steps be taken to give the manufacture of these shells, on the system which has been lately adopted by Mr Nordenfelt (sic), a fair trial. To this end Mr Nordenfelt might be asked to supply the rough castings for 500 common and 500 shrapnel shell for the 12-pounder B.L. Gun, and these might then be submitted (to) a comparative examination and trial, either at home or in India, with the rough forging of the same nature, as supplied to Woolwich. The shell could be finished at Cossipore, if comparative trial was carried out in India. If the result should be in favour of Mr Nordenfelt's system, I would urge that no time should be lost in introducing the system of casting on the Mitis principle into Ordnance Factories in India, and that arrangements be entered into with Mr Nordenfelt in regard to the patent rights, and the introduction of the manufacture into India.[23]

The Government of India finally echoed the voice of the Director General of Ordnance in its letter to the Secretary of State on 20 September 1887. It stated that a sufficient quantity of scrap iron suitable for the production of Mitis castings would be available in India and suggested proper testing of the various articles manufactured by this process before their final adoption in India, where "the climatic changes (we)re far greater than England".[24] There must have been some trials and experiments on this method of production, and some pressure groups might be operating for and against its adoption. They were finally dismissed in February 1888 when the Mitis castings submitted for trial in the Royal Arsenal had turned out full of blow holes, so "as to render it inadvisable to send out any for trial in India at present".[25] But the significance of the development was that

Cossipore became the hotbed of a technological debate even before the beginning of its steel production.

It was by the end of 1890 that the Director General of Ordnance submitted the final proposal to the Government of India for the manufacture of steel in the FSF, Cossipore. The general introduction of breech-loading ordnance in India had necessitated the provision of steel projectiles, thereby providing for the manufacture of steel shells as used for breech-loading guns. As the Director General elaborated in his letter dated 2 December 1890:

> A considerable part of the plant for the manufacture of these shells already exists in Cossipore, but in order to utilise it to the best advantage, the manufacturing power of the factory must now be supplemented by the appliances and establishments required for the production of steel, both cast and forged. *If this is not done, the only shell factory in India will no longer be self-contained.*
>
> The supply of ammunition for guns is also not the only consideration. Steel enters very largely into the construction of the carriages for field artillery and of all mountings for heavy guns, and owing to there being at the present time no source of supply in India, the smallest repair in cast steel cannot be executed without preferring a demand on England for the necessary material ... I am of the opinion that the present requirements will be met by its introduction into the Cossipore Factory, in which the question for the manufacture for steel shell for guns which are actually under supply, has to be considered first of all, and the source of supply of the necessary components determined. If the manufacture of cast and forged steel should prove successful, of which there is no reasonable doubt, most valuable assistance then could be given to the gun carriage factories in the supply of steel components and material, and there is no limit probably to the eventual requirements of the Ordnance Department as regards the supply of different articles of war materials, machinery and miscellaneous castings ... .
>
> For the production of cast steel, I propose that a 3-ton furnace on the Siemens-Martin open-hearth principle be erected. This is the system most generally adopted in England and is considered to give the best results, while being easier of manipulation, and less liable to derangement than others ... .[26]

Certain findings on the product cost of the proposed manufacture were also interesting. The cost of cast steel made at Cossipore was estimated at £ 22.10 per ton as against £ 25.13 which was approximately the cost of the Royal Laboratory, exclusive of depreciation charges. The cost of the forged steel projectiles was estimated at Cossipore on the first out-turn with imported bar steel as £ 53 per ton against £ 56 and £ 58 per ton in England for the unfinished components of 12-pounder shrapnel and common shells. The Director General noted that the main reason for the reduced cost of both cast and forged steel made in India was the cheap rate of labour obtainable here compared to that paid in England. Most interestingly, he added:

> But it is no secret that a few firms in the trade, having had hitherto almost a monopoly in the supply of steel forgings, have demanded a high price which has been further maintained by the inability of the trade generally to compete for the comparatively small orders such as shells with their ever varying forms and specifications.[27]

It is important to note that the Director General of Ordinance had already received a memorandum dated 2 October 1890 and submitted by the Superintendent, India Store Department, which questioned the Cossipore venture on grounds of production and prices, thereby suggesting a delay in the manufacture. He became furious. He found 'no valid objection' in the memorandum and added in his letter to the Military Department on 31 December 1890:

> At present ... the Ordnance Department is absolutely and deplorably dependent on England for all articles of steel. The pivot of a platform cannot be replaced without an order on England. It is impossible that such a state of things can last long if the policy of independence for India is to be maintained, and I beg once more to urge that the manufacture of steel shall be undertaken with as little delay as possible.[28]

The Government of India finally approved the project for the manufacture of steel in Cossipore, and Lord Lansdown, the Governor General and Viceroy of India, in his letter to the Secretary of State for India on 25 February 1891 asked for a sanction of Rs 1,25,000 for the purpose.[29] Of this amount, the cost of importing and setting up the new plant was about Rs 1,20,000 (or 9,000 Pounds @ 1s 6d per rupee). There would be a recurring extra charge of Rs 5,700 per annum to meet the additional establishment after the commencement of the manufacture. The final approval came from the Secretary of the State in May 1891: "In view of the advantage of India being, in time of war, independent of this country in regard to an efficient supply of artillery ammunition, and in view of the opinion expressed by the Director General of Ordinance in India that steel can be made in India at such rates that the cost of projectiles and other articles made from it will compare favourably with the cost your Government of similar articles manufactured in the

Government factories in England, I have resolved to sanction your several proposal."[30] Captain Mahon, already deputed in England, was authorised to supervise over the supply of the necessary plant of machinery from that country for Cossipore. Mr Lomas, Foreman of the Foundry and Mr Kidd, Foreman of the Shell Factory, along with the four men of the Carnatic Artificer Company working in the Cossipore Factory were sent to England for necessary training and instruction. The original plan was even to bring from England the shed to contain the plant of the machinery for the steel foundry at Cossipore,[31] but it was subsequently changed and it was decided that the building to contain the new plant of machinery should be made at Cossipore.[32]

New Technology and the Manufacture of Field Guns in Cossipore

In the wake of the French and German initiatives by the close of the nineteenth century, the European armies came to have a new type of field gun based on improved technology. The accuracy and rapidity of fire under this new technology revolutionised gunnery and marked the advent of quick-firing guns.[33] The British artillery underwent major change in the early years of the twentieth century, and the introduction of the manufacture of field guns in Cossipore was a direct response to it. The very name of the factory was changed to Gun and Shell Factory, Cossipore and the new work commenced in 1905. The technological question was very important: the conditions under which it was generated in Europe, then adopted in England and finally fostered in India deserve our special attention.

In Europe in the late nineteenth century, the rate of fire of artillery had attained the upper limit set by the recoil of a field gun. After each firing, the gun leaped backward with the force of the blast, and the gunners had to push the whole gun back into position and aim it again after each round. In 1896, two French Officers named Captain Deport and Sainte Claire Deville tried to solve this problem. "They presented a design, which the French Army adopted, for a field gun carriage that remained static while firing, only the barrel itself slid backward, along a track in the mounting. The shock was taken up by an intricate brake, a piston cylinder mounted parallel to the barrel, which used the action of the recoil to compress fluid. The compression and re-expansion of the brake first absorbed the recoil of the barrel and then pushed it forward again into exactly the position from which it had started."[34] The famous French 75mm gun that was introduced with these features in 1897 made the field artillery of other countries obsolete. "The old guns were theoretically capable of firing up to 5 rounds per minute, though precise aiming each time took longer. By contrast, the Q.F. French 75 mm gun could fire 20 rounds per minute".[35] To compete with the French model, the Germans designed their 77mm gun, which was given spring recoil. The British followed the German principle.

The technological change was very much in the process when the first proposal for the manufacture of field guns in India came from the Director General of Ordnance in India in February 1903. Naturally, Cossipore was given a position of prime importance in this manufacture. "The manufacture of gun components, breech screws, carrier rings etc., (wa)s well-established at Cossipore on a scale to the present demands, and the production of necessary quality of steel, the forging, the treatment of the steel (oil hardening and tempering) and finally finishing with necessary accuracy ... ".[36] Cossipore's outturn was fully satisfying as it could rigidly follow all Home specifications and tests in India. Major General C.H. Scott, the Director General of Ordnance further reported to the Military Department:

> With the well-trained supervision that we get from English works and the careful proof that is applied to the outturn, there is no doubt of the success that will attend the manufacture of guns in India. The work is most suitable to the native workmen who excel in the straightforward work like boring and turning, and as the majority of the work on a gun is of this class, success and economy will certainly result.[37]

There were certain problems and prospects also. Cossipore was not yet in a position to cast large ingots required for gun work, hence certain import was necessary. As the quality of the material was the most important factor in the serviceability of a gun, rough forgings might be obtained from England, but the fact was that it would be cheaper to make them in India. Scott, therefore, recommended that steel should be brought from England in the form of billets, which themselves were forged from still larger ingots. "If in future we are able to undertake the casting and forging of the heavy ingots ourselves, a much greater saving will be made."[38]

Situation had been changing rapidly and the Government of India took a major decision to begin the manufacture of field guns in India. In March 1903, Lord Curzon, the Governor General and Viceroy of India, wrote to the Secretary of State with much prudence and confidence.

... we have of recent years pursued ... the principle of making India independent of assistance from home in the matter of war material ... we now propose to extend the operations of the existing factory at Cossipore and to include in them the manufacture of guns. The present appears to be a favourable moment for inaugurating such a scheme. We understand that it is in contemplation to rearm the field artillery at home with a quick-firing gun. A similar re-armament of batteries on the Indian subcontinent must follow ... we anticipate that with the skilled supervision, which is at our disposal, there will be no difficulty in manufacturing field guns of the most modern type at Cossipore ... .[39]

All the problems were not over, and there were of course questions from certain authorities or pressure groups from England. One such question was about the "interchangeability of the pattern of guns made in India and at home". As the Superintendent of the Royal Gun Factory pointed out, there would not be any difficulty in manufacturing field guns in India at Cossipore "if the work done (wa)s to be strictly interchangeable with that turned out in this country by Ordnance factories and trade, models, gauges etc., will have to be supplied from home to ensure interchangeability, *it being practically impossible to get it without*".[40] The War Office also supported this opinion which "should be absolutely insisted upon" and noted with approbation that even Lord George Hamilton confirmed this. But one interesting point was highlighted in War Office's letter to India Office on 25 June 1903:

> Though the manufacture of field guns in India will undoubtedly be to some extent prejudicial to the Ordnance factories and to private manufacturing firms at home, Mr Brodrick is aware that important advantages will be gained by extending the sources of supply of warlike materials within the Empire, and by making India more independent than she now is. He considers, therefore, that the proposal should be given effect to.[41]

In July 1904, Ampthill, the Governor General of India, reported to the Secretary of State on the more improved version of the project for the manufacture of field guns in India.[42] By this time, the plan of Major Renny, the last Superintendent (June 1899–October 1905) of the FSF, Cossipore for more expensive and efficient machinery so as to accommodate the manufacture of 5-inch B.L. Gun had been approved by the Secretary of State. The sanctioned amount for the supply of machinery recommended by Major Renny was up to a cost of #18,375. Ampthill was answering three queries of the Secretary of State, namely, on the use of steel billets or forgings, the number of skilled mechanics to be required, and the type of material to be supplied from England for the manufacture of field guns. His clarification gives us an idea of the technology used in this production in Cossipore in the early years of the twentieth century. The manufacture would be from steel billets in preference to forgings, unless the latter could be supplied at rates more economical to manufacture from them. Major Renny's advice would be generally accepted on this matter. Secondly, one Assistant Foreman had already been engaged for the forge department, but two more would be needed for the gun plant and also a Foreman for the examination branch. The latter should be a trained mechanic qualified to inspect guns and gun-fittings and also to pass signaling apparatus and telescopes. It is important to note that the repairs of telescopes and binoculars as well as the manufacture of signaling apparatus in Cossipore had been approved as early as April 1903 by the Director General of Ordnance in India.[43] Finally, the materials required from England, as a first supply for the proposed manufacture would be forgings for sixty guns and billets for forty guns.[44] All of these 100 guns would be Q.F. Guns, of which 95 guns would be 12 ½ pounders and 5 guns would be 18 ½ pounders. Incidentally, this was the total number of Q.F. Guns supplied to India by the War Office in 1904–05, and the proposed manufacture was for 100 per cent substitution. Major Renny, who was to be given great credit for the most effective plant thus raised left the Factory in October 1905 before the manufacture was actually established and his successor, Major Bell, the first Superintendent (October 1905–July 1912) of Gun and Shell Factory, Cossipore "was fortunately able after a most anxious and trying time, to turn out guns equal in all respects to the best home manufacture".[45] It was cheaper also.[46]

Thus, technology was a double-edged affair in Cossipore as far as ordnance production and policy was concerned. While the very limitation in its application with adverse bearing on production was the direct consequence of the policy pursued at the higher level under conditions of colonial dependence, the highly competitive world of the technology of warfare forced partial modification of the defence policy of India, thereby limiting colonial dependence on this matter. However, the basic dichotomy was not fully resolved before Independence, and Cossipore's history between and during the two World Wars was a reflection of that.

Notes

1. See Michael Adas, *Machines as the Measure of Man: Science, Technology and Ideologies of Western Dominance*. Delhi: Oxford University Press, 1990, p. 367.

2. See Major General K.C. Mehra, *A History of the Army Ordnance Corps 1775–1974: A Saga of Two Centuries*. Allahabad: Wheeler Publishing, 1980, p. 127.
3. See Michael Howard, *The Franco-Prussian War: The German Invasion of France, 1870–1871*. London: Routledge, 1961, reprint 1981, p. 5.
4. See Correlle Barnett, *Britain and Her Army, 1509–1970: A Military, Political and Social Survey*. London: Penguin, 1970, p. 342.
5. Ibid., p. 328.
6. Martin van Craweld, *Technology and War: From 2000 BC to the Present*. London: Collier Macmillan, 1991, p. 297.
7. J.F.C. Fuller, *Armament and History: The Influence of Armament on History from the Dawn of Classical Warfare to the End of the Second World War*. New York: Vintage Books, 1945, reprint 1998, p. 31.
8. From the Inspector General of Ordnance and Magazines to the Military Department, 20 June 1860, para 2 in PMD, August 1860, No 4, National Archives of India (hereafter NAI.).
9. From Northbrook to the Secretary of State for India, 4 August 1873 in *Military Despatches to the Secretary of State for 1873*, Letter No. 161 of 1873, NAI.
10. See *Appendices to the Report of the Special Ordnance Commission appointed by the Order of the Government of India to enquire into the System of Ordnance Supply in India*. Calcutta, 1875, New Delhi: Central Secretariat Library, p. 328.
11. See W.A. Johnson, *The Steel Industry of India*. Cambridge, Mass: Harvard University Press, 1966.
12. See Morris D. Morris, 'The Growth of Large-scale Industry to 1947', in Dharma Kumar (ed.), *The Cambridge Economic History of India, Volume II, c. 1757–c. 1970*. Cambridge: Cambridge University Press, 1982, p. 591.
13. Ibid., p. 584.
14. See George Jessop of Jessop and Company to Holt Mackenzie, Secretary to the Government's Territorial Department, May 1825, quoted in Arun Bandopadhyay, *The Story of Jessop*. Calcutta: Motilal Banarsidass, 1988, p. 26.
15. Morris D. Morris in Dharma Kumar (ed.), op. cit., p. 585.
16. Ibid.
17. Richard A. Preston and Sidney F. Wise, *Men in Arms: A History of Warfare and its Interrelationship with Western Society*. New York: Preager, 1956, reprint 1979, p. 243.
18. From Captain Mahon, Officiating Supdt., FSF, Cossipore to IG of Ord., Bengal, 5 July 1887 in the Manufacture of Steel in India, L/MIL/7/13792, Oriental and India Office Library (hereafter OIOL).
19. Quoted in ibid.
20. Ibid.
21. Colonel Shea to the Deputy IG of Ord., Bengal Circle, 23 July 1887 in ibid. Italics are added.

22. From R.F. Lowis, Director General of Ordnance in India to the Secretary to GOI, Military Department, 18 August 1887 para 4, in ibid.
23. From R.Wace to the Military Secretary, India Office, London, 21 September 1887 in ibid.
24. See extract from a Despatch of GOI to the Secretary of State, 20 September 1887, No. 163, Para 14 in ibid.
25. From the Ordnance Consulting Officer in India to the India Office, London, 24 February 1888 in ibid.
26. From Maj. Gen. A. Walker, Director-General of Ordnance in India to the GOI in the Military Department, 2 December 1890, paras 3–5 in ibid. Italics are added.
27. Ibid.
28. From the DG of Ord., to the Military Department, 31 December 1890, in ibid.
29. From the GG of India to the Secretary of State, 25 February 1891, para 4, in ibid.
30. From the Secretary of State to the GG of India, 14 May 1891, Military Letter No. 70, Para 2, in ibid.
31. See extract of the letter from the GOI to the Secretary of State, 25 November 1891, Para 29, quoted in ibid.
32. See extract of the letter from the GOI to the Secretary of State, 20 January 1892, Para 26, quoted in ibid.
33. See Mehra, op. cit., p. 130.
34. See David G. Herrmann, *The Arming of Europe and the Making of the First World War*. New Jersey: Princeton, 1996, p. 17.
35. Ibid.
36. From Scott, the Director General of Ordnance in India to the Military Department, 7 February 1903 in Manufacture of Steel and Guns in India, L/MIL/7/13796, OIOL.
37. Ibid.
38. Ibid. He continued: "The furnaces and tanks for oil hardening, annealing and heating for shrinkage, will occupy the shed now covering the brass furnaces. Such furnaces as are required will all be made locally from designs that can be got in England. The oil tanks are already in use."
39. From the GG of India to the Secretary of State, 26 March 1903, Military Department No. 55 of 1903, in ibid.
40. From the Superintendent, Royal Gun Factory, to the Chief Superintendent, Ordnance Factories, England, n.d., as Annexure I of the Letter from War Office to India Office, 25 June 1903, in ibid.
41. From the War Office to the India Office, 25 June 1903, as Enclosure 2 in L/MIL/7/13796, OIOL.
42. From the GG of India to the Secretary of State, 28 July 1904 in ibid.

43. From the DG of Ord. to the Superintendent, FSF, Cossipore, 16 April 1903 in ibid. "Please include in your final proposals in connection with the manufacture of guns at Cossipore the provision of a trained civilian mechanic for the examination branch at Cossipore who will also be engaged in the examination and passing of signalling apparatus and telescopes." Para 2.
44. "The materials for manufacture required from home as a first supply will be forgings for sixty guns and billets for forty guns or if on Major Renny's advice billets are not to be used, forgings for 100 guns should be supplied." See GG of India to Secretary of State, 28 July 1904, Para 3, in ibid.
45. A Brief History of the Government Gun and Shell Factory at Cossipore, Calcutta, 1910, p. 10.
46. Ibid.

21 Jawaharlal Nehru and Science and Technology

The 1958 Scientific Policy Resolution Reconsidered

Nasir Tyabji

Jawaharlal Nehru is, of course, identified in the public mind with the tireless advocacy of the intrinsic link between science and development. In the early years after independence this association was generally perceived in a positive spirit, as was the notion of India's transition to modernity, with which this linkage was generally connected. However, sometime in the late 1970s, a current of disenchantment with the ecological, social and cultural effects of development projects set in. This was because of the chasm between those sections of society that benefited from the projects and those that bore the direct costs incurred in the construction of these projects, in terms of the disruption of their culture and means of livelihood. The association that these projects in themselves were perceived to have with the Nehruvian vision (rather than the authoritarian manner of their implementation) led to an increasing alienation of some sections of the intelligentsia from this vision. This period, coinciding with the first stirrings of the liberalisation process which was associated with the 1976 budget, led to a more general questioning of Nehru's legacy, in the doctrines of Nehruvian socialism.

Science also played a role in the second feature of the Nehruvian vision, which consisted of the process of building a secular society. Here it was the creation, on a mass scale, of what came later to be termed as the 'scientific temper' that was the aim to be achieved. This, too, became the subject of violent controversy in the years leading up to the upsurge in communal forms of nationalism in the mid-1980s. With this, Nehruvianism itself faced an eclipse which was accelerated by the ascendancy of the economic ideology underpinning the unfettered process of economic liberalisation initiated by the 1991 budget. It is only now, almost thirty years after the criticism began, that a new appraisal of Nehruvianism is beginning to take place, the defining feature of this new phase being the political fallout of the 2004 elections. This fallout has created the space which allows for a reorientation from the purely defensive strategies which had necessarily to be adopted during the period when the forces of communal nationalism were directly in control of the government.

In the meanwhile a major issue that has entered the centre space of the Indian political and social arena is that of empowerment. This issue has expanded in scope from the constitutionally ordained affirmative action in favour of the dalits and the tribal people to the other, largely peasant communities which fall within the category of the other backward castes. To this must be added the women's question. Although this last had been an integral part of the national movement, not only as a part of Gandhian thought, but was actually the subject of deliberation by a sub-committee of the National Planning Committee, it achieved new salience in the 1970s with the growth of the women's movement.

The question is then of the degree to which Nehru's vision of the role of science and technology is able to encapsulate these new aspirations, rather for a form of development that incorporates these aspirations. It is a well-known characteristic of Jawaharlal Nehru that many of his statements of intention were made as an individual, and amounted to statements of personal vision. Frequently, they were not operationalised into Congress strategy before Independence or formed a part of government policy after independence. This was also true of his observations on the role of science, often made while inaugurating sessions of the Indian Science Congress. However, the 1958 Science Policy Resolution stands on a different footing.[1] As a statement of government policy, it had to be processed by the relevant agencies of the government and represented a collective view of the role of science in the developmental process. However, even though it was drafted by Homi Bhabha, it had the imprint of Nehru's personal vision. It is interesting to return to this document to answer the question of the durability of Nehru's vision on science and technology, almost 50 years after it was published.

The Scientific Policy Resolution is, unlike the Industrial Policy Resolution of 1956, a short document consisting of seven paragraphs. The first paragraph itself shows an unusual perception of the determinants of development when it states:

> The key to national prosperity, apart from the spirit of the people, lies, in the modern age, in the effective combination of three factors, technology, raw materials and capital, of which the first is perhaps the most important, since the creation and adoption of new scientific techniques can, in fact, make up for the deficiency in natural resources, and reduce the demands on capital. But technology can only grow out of the study of science and its application.

The unusual feature here is, of course, the reference to the spirit of the people as one of the key factors. The reference here is clearly to the imperative of arousing the enthusiasm of the bulk of the people, for in the absence of their active participation, the process of development would be slowed down if not thwarted.[2]

The critical nature of the impact of science in altering the consciousness of the people continues to be emphasised in the second paragraph:

> The dominating feature of the contemporary world is the intense cultivation of science on a large scale, and its application to meet a country's requirements. It is this, which, for the first time in man's history, has given to the common man in countries advanced in science, a standard of living and social and cultural amenities, which were once confined to a very small privileged minority of the population.

The critical point to be noted here is the use of the word 'cultivated' in association with science, as compared to the more mundane reference to the application of science. This surely implies that science is first to be imbibed, and then be actively applied, not only by scientific agencies, but also by the bulk of the population.

The second paragraph continues:

> Science has led to the growth and diffusion of culture to an extent never possible before. It has not only radically altered man's material environment, but, what is of still deeper significance, it has provided new tools of thought and has extended man's mental horizon. It has thus influenced even the basic values of life, and given to civilization a new vitality and a new dynamism.

It is thus the spread of the scientific temper which not only leads to the spread of culture but also invigorates pre-existing cultures, providing for a more inclusivist process of growth. Indeed, the spread of the scientific temper is critical to the growth of the productive forces and the provision of economic, social and cultural amenities that this provides.

> It is only through the scientific approach and method and the use of scientific knowledge that reasonable material and cultural amenities and services can be provided for every member of the community, and it is out of recognition of this possibility that the idea of a welfare state has grown. It is characteristic of the present world that the progress towards the practical realisation of a welfare state differs widely from country to country in direct relation to the extent of industrialisation and the effort and resources applied in the pursuit of science.

This formulation, that the progress towards the achievement of a welfare state depends on the degree of industrialisation achieved, marks the transition in the policy resolution from a discussion of science to that of technology. Before turning to the question of technology, it may be emphasised here that at the conceptual level, the question of the role of science has been expounded in an extremely sophisticated way. The familiar problems that plague the practice of science in India are not due to the vision which underlay it. These problems, which have prevented science from fulfilling the developmental goals so eloquently described in the Scientific Policy Resolution, are due to three main reasons. The first is the creation of structures of scientific research (CSIR, the Department of Atomic Energy, and the Defence Research and Development Organisation) outside the framework of the university system. Secondly, this weakening of the university research effort caused by the drainage of scientists from the universities into organisations offering comparatively better career prospects, has been compounded by the concentration of research funds precisely on these three organisations. Finally, this concentration has led to the domination of not only 'big science', but more corrosively, on the overwhelming emphasis to the problems of the defence of the state, or at least, to the formulation of problems as perceived by the state. The net result has been that the option of utilising the transformative powers of science potentially realisable by relating science to the aspirations of the people was consciously rejected. Campaigns for the inculcation of the scientific temper, for instance, had sporadic existence rather than forming a component part of the science activity itself.

On the question of technology, the Scientific Policy Resolution descends to a disappointingly mundane plane. In the fourth paragraph it is stated that:

> The wealth and prosperity of a nation depend(s) on the effective utilisation of its human and material resources through industrialisation. The use of human material for industrialisation demands

its education in science and training in technical skills. Industry opens up possibilities of greater fulfillment for the individual. India's enormous resources of manpower can only become an asset in the modern world when trained and educated.

The key feature to note in this paragraph, the only mention of the human factor in this part of the resolution, is the instrumentalist view not only of the role of scientific and technical training, but of the human being itself. It is reduced to an input to production, qualitatively at the same level as the material resources. This is, of course, consistent with the homogenised approach to factors of production in the neo-classical analysis of economics, but represents a surprising change of gear within the Scientific Policy Resolution. It also signifies a conception of technology, the area of knowledge underlying the production process, and thus of industrialisation, which is exclusively 'engineering' in conception.

From the work of Frederick Taylor dating to a century ago, it has been made evident that the human process of work, or the labour process, is an intrinsic part of the production process. Production can only take place with the combination of the labour process with the physical process of the conversion of the raw materials into finished commodities. Following Taylor's research, the explicit understanding of the need to incorporate the labour process into manufacture was exemplified by the Fordist innovation of the production line and this understanding has continued with the work procedures within contemporary Toyotaism. The approach to understanding the labour process, however, is from a managerial perspective, basing itself on the need to control the process rather than viewing it as an opportunity to empower labour, though Toyotaism does mark an advance over Fordist production methods in this respect. However, the achievement of collective bargaining rights by the labour movement in advanced industrial economies has also brought the concept of the labour process into general recognition though, again, this is from a managerial perspective.

Labour relations in India did not evolve into Taylorism in the pre-independence period.[3] Arising from this, the concept of true collective bargaining has not been recognised in law, even after independence. Following the adoption of the model of arbitration, initially introduced by Gandhi into the Ahmedabad industrial milieu, this approach was first extended by law to the whole of Bombay Presidency by the late 1930s, and to the rest of India by the time of independence.[4] Underlying Indian labour relations legislation is the view that the State must be in a position to arbitrate. Provision for the enduring role of the state has distorted the process of negotiation between management and workers and effectively prevented the true development of collective bargaining.[5]

Unlike the situation with regard to science, therefore, where the Scientific Policy Resolution could start *de novo*, technology was an area where existing forms of the social relations within production prevented the development of a conception of the work process, particularly that conception which was empowering for workers. The entire history of post-independence developments militated against the recognition, let alone the operationalising, of such a notion of the relationship between workers and management. The Congress inspired split in the labour movement, signified by the breaking away of the Indian National Trade Union Congress (INTUC) from the All India Trade Union Congress (AITUC) in the period immediately preceding independence, while ensuring that the Congress retained a base within the labour movement, was fatal for the creation of a collective bargaining framework. This development, largely the handiwork of the trade union leaders within the Congress, supported by Sardar Patel, probably was not within Nehru's power to control. However, what is surprising is the fact that when, a few years after independence, V.V. Giri, Labour Minister in Nehru's Cabinet, wished to introduce a Bill amending the Industrial Disputes Bill the matter was allowed to drag on for several months. Finally, Nehru suggested that Giri discuss the amending bill with T.T. Krishnamachari, Minister for Commerce and Industry, an unusual person to select, given that given his portfolio, he would represent the employers of labour. Eventually, the Bill was allowed to lapse after Giri left the Ministry. Subsequent Labour Ministers did not show equal determination to introduce formal collective bargaining.

Although the connection would be difficult to establish without further detailed research, it seems plausible to conclude that the sharp contrast between the democratic and sophisticated associations drawn between the role of the human being and the role of science in development are missing in the discussion of industrialisation because of the historical events outlined above. The organisation of science presented undoubted problems in preventing the growth of dysfunctional hierarchies, and the imperatives of perceived state priorities was to prevent science playing a liberatory role, in the overall scheme, science was not central to the political economy. The organisation of industry, on the other hand, was a defining feature of this political economy, and it was not possible to postulate a social arrangement within the production process which was at odds with the prevailing hierarchy.

The Nehruvian vision of science and technology as embodied in the Scientific Policy Resolution was a far-seeing document of statement of intent. As far a science is concerned, though the actual practice of science was far removed from the vision, this hiatus is capable of being corrected if the social forces presently excluded from, or marginalised by, the development process are enabled to assert themselves. However, the perspective on technology is lacking in complementary depth. It is the premise of this paper that the mechanical conception of the human role in production stems from the actual social role to which workers are confined by the existing industrial relations system. In this case, the postulates of the Scientific Policy Resolution themselves require to be restated in accordance with the social aspirations of the time.

Notes

1. Government of India, Scientific Policy Resolution, March 4, 1958, No. 131/CF/57, New Delhi.
2. At another, more technical level, the resolution shows an obsolete notion of the relationship between science and technology in the assertion that "... technology can only grow out of the study of science and its application." This representation of a linear relationship between science and technology, later characterised as the ladder approach was replaced with a more nuanced understanding when Toyotaism introduced the cyclical relationship.
3. See, for example, Arup Kumar Sen, "Mode of Labour Control in Colonial India", *Economic and Political Weekly*, Vol. XXXVII, 38, 2002, pp. 3956–66.
4. Sujata Patel, *The Making of Industrial Relations*. New Delhi: Oxford University Press, 1987.
5. Debi S. Saini, "Labour Legislation and Social Justice: Rhetoric and Reality", *Economic and Political Weekly*, Vol. XXXIV, 1999, L32–L40.

Contributors

R. Balasubramaniam, Professor, Department of Metallurgical Engineering and Material Sciences, Indian Institute of Technology, Kanpur.

Arun Bandopadhyay, Professor, Department of History, Kolkata University, Kolkata.

P.K. Basu, Professor, Department of Philosophy, University of Hyderabad, Hyderabad.

Nupur Dasgupta, Professor, Department of History, Jadavpur University, Kolkata.

Suchandra Ghosh, Associate Professor, Department of History, Kolkata University, Kolkata.

S. Irfan Habib, Professor, National Institute of Science, Technology and Development Studies, New Delhi.

Najaf Haider, Associate Professor, Centre for Historical Studies, Jawaharlal Nehru University, New Delhi.

Rattan Lal Hangloo, Professor, Department of History, University of Hyderabad, Hyderabad.

E. Haribabu, Professor, Department of Sociology, University of Hyderabad, Hyderabad.

S. Jay Kishan, Associate Professor, SLNSA Oriental Degree College, Dharmapuria, Andhra Pradesh.

Jaya Menon, Associate Professor, Centre for Advanced Study, Aligarh Muslim University, Aligarh.

Debasmita Patra, Research Scholar, Department of Sociology, University of Hyderabad, Hyderabad.

Dhruv Raina, Professor, Centre For Science Policy, Jawaharlal Nehru University, New Delhi.

M.N. Rajesh, Assistant Professor, Department of History, University of Hyderabad, Hyderabad.

M. Ramakrishna, Ph.D. Research Scholar, Department of History, University of Hyderabad, Hyderabad.

Vijaya Ramaswamy, Professor, Centre for Historical Studies, Jawaharlal Nehru University, New Delhi.

Syed Ali Nadeem Rezavi, Associate Professor, Centre for Advanced Study in History, Aligarh Muslim University, Aligarh.

K.N. Sethi, Associate Professor, University of Sambalpur, Sambalpur, Orissa.

Surinder Singh, Professor, Department of History, Punjab University, Chandigarh.

Arun Sonakia, Senior Scientist, Geological Survey of India, Nagpur, Maharashtra (formerly Director, Geological Survey of India, Nagpur).

Sharda Srinivasan, Professor and Senior Scientist, Indian Institute of Science, Bangalore.

S.S. Subodh, Associate Professor, Department of History, University of Hyderabad, Hyderabad.

Nasir Tayabji, Professor and Director, Centre for Jawaharlal Nehru Studies, Jamia Millia Islamia, New Delhi.